The Arctic: A Very Short Introduction

VERY SHORT INTRODUCTIONS are for anyone wanting a stimulating and accessible way into a new subject. They are written by experts, and have been translated into more than 45 different languages.

The series began in 1995, and now covers a wide variety of topics in every discipline. The VSI library currently contains over 650 volumes—a Very Short Introduction to everything from Psychology and Philosophy of Science to American History and Relativity—and continues to grow in every subject area.

Very Short Introductions available now:

Available soon:

For more information visit our website

www.oup.com/vsi/

Klaus Dodds and Jamie Woodward

THE ARCTIC

A Very Short Introduction

OXFORD
UNIVERSITY PRESS

OXFORD
UNIVERSITY PRESS

Great Clarendon Street, Oxford, OX2 6DP,
United Kingdom

Oxford University Press is a department of the University of Oxford.
It furthers the University's objective of excellence in research, scholarship,
and education by publishing worldwide. Oxford is a registered trade mark of
Oxford University Press in the UK and in certain other countries

Published in the United States of America by Oxford University Press
198 Madison Avenue, New York, NY 10016, United States of America

British Library Cataloguing in Publication Data

Data available

Library of Congress Control Number: 2021941451

ISBN 978-0-19-881928-8

Printed in Great Britain by
Ashford Colour Press Ltd, Gosport, Hampshire

Acknowledgements

The world changed profoundly during the writing of this book.
On 2 March 2020 we met at the Royal Geographical Society in
London to map out the final stages of this project. We even
booked the *Arctic Room* where we reviewed progress and set some
ambitious writing deadlines. It was a good meeting and we made
plans to see the British Museum's exhibition on *Arctic Culture and
Climate* at the end of the year. We agreed to take stock in early
April when we would be back at the RGS for a meeting of the
Society's trustees. On 23 March the UK went into the first
lockdown of the Covid 19 Pandemic and the RGS closed its doors.
We have not met in person since—the rest of this collaboration
took place via Zoom and email. We did not make it to the BM.

The full impacts of the Covid 19 pandemic on Arctic communities
will not be known for some time, but lessons are being learned
about how to prevent the spread of the virus to more remote parts
of the Arctic where access to healthcare is limited. Just a few days
ago officials in Greenland sealed off Nuuk amid concerns around
an outbreak of Covid. The Arctic presents distinctive challenges to
the vaccination effort but there are hopeful signs. As we write
these acknowledgements in June 2021 about a quarter of
Greenland's residents have been vaccinated.

Many people have helped us to write this book. We would like to thank Jenny Nugee and Latha Menon at Oxford University Press for their editorial support and expert guidance. We also thank Nivedha Vinayagamurthy and her team for overseeing the production of this VSI. As part of the editorial process, we were fortunate to have Professor Mark Maslin as an expert reader and his insightful suggestions were gratefully received. We have also benefited immensely from our circle of academic colleagues and friends who devote their professional lives to making sense of the Arctic and environmental change. KD gratefully acknowledges the funding support of the Leverhulme Trust in the form of a Major Research Fellowship (2017–2020) that examined the contemporary geopolitics of the Arctic. JW is grateful to The University of Manchester for sabbatical leave when a good deal of this book was written. We would also like to thank the National Health Service (NHS)—we have both been vaccinated twice.

We would like to thank our families for their love and support. The RGS is a wonderful place to meet and talk about the Arctic—you may pass a bust of the indefatigable Lady Jane Franklin on the ground floor. We hope to be back there soon.

Contents

List of illustrations

Hughes, Philip & Gibbard, Philip & Ehlers, Jürgen. (2013). Timing of glaciation during the last glacial cycle: Evaluating the concept of a global 'Last Glacial Maximum' (LGM). Earth-Science Reviews. 125. 171–198. 10.1016/j.earscirev.2013.07.003.

Chapter 1
The Arctic world

The Earth is warming. As humans have continued to add greenhouse gases to the atmosphere, every decade since 1980 has been warmer than the last. But the warming of our world is far from uniform. The Arctic is warming at least twice as rapidly as the global average and some parts of the far north are warming three to four times faster. This phenomenon—the product of a cascade of feedbacks that magnify the warming trend—is known as *Arctic amplification*. Since about 1980, the warming trajectory in the Arctic has been much steeper than that of the rest of the planet (Figure 1). The Arctic has already far exceeded the goal of the 2015 Paris Agreement to limit warming to 1.5°C compared to pre-industrial levels.

We finished writing this book in early 2021, as the various environmental datasets for 2020 were compiled. They showed very clearly that 2020 was a milestone year for the far north—its summer was the hottest ever recorded. On 20 June 2020, in the midst of an unprecedented heatwave, air temperatures in the Siberian town of Verkhoyansk (67.5° N) reached 38°C (100.4°F)—a record high temperature inside the Arctic Circle. Weather conditions once considered extreme are becoming the norm. Rain is replacing snow. Wildfires are more frequent, more extensive, more intense. The Arctic is changing. Baselines have

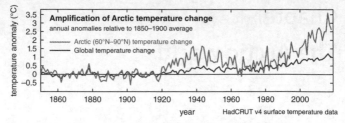

1. **Global and Arctic temperature change since 1850.**

shifted. Permafrost thaw and coastal erosion are accelerating. Summer lightning strikes are now ten times more numerous than they were just a decade ago because warming favours the formation of convective storm clouds. Some climate scientists believe that the Arctic has already shifted to a new climate regime.

In a little over a generation, sea ice extent in the Arctic at the end of summer has fallen by half. Sea ice is part of the reflective shield that helps keep the Arctic cool. The loss of glacial ice from Greenland has accelerated so quickly in the past decade it is now tracking the worst-case scenarios set out in the 2014 Intergovernmental Panel on Climate Change (IPCC) Assessment Report. This acceleration has taken the scientific community by surprise. The Arctic is now exceeding climate change predictions by decades—it will feature prominently in the Sixth Assessment Report (AR6) of the IPCC due in 2022, especially in relation to climate change impacts, adaptation, and vulnerability. A vast human-induced experiment in rapid climate and ecological change is playing out at the top of the world. These environmental changes have consequences for all of us.

Against this background, this very short introduction aims to capture the aspirations, opportunities, and challenges facing those who live and work in the Arctic. It will also account for global interest in the northern edges of our planet—environmental, social, geopolitical, and economic. It is a complex story that reveals different worldviews and competing interests and priorities, as well as recognition that the Arctic is richly complex in its landscapes, ecosystems, peoples, cultures, and resources. But as we write at the beginning of a new decade, it is also one that bears the imprint of timing. Could the summer of 2020 be representative of this new Arctic? Since both climate and ecosystems are changing so rapidly in the Arctic, past definitions of the region are being stress-tested to breaking point.

Defining the Arctic

In physical geography textbooks, the Arctic is traditionally defined as the land, sea, and ice lying north of the Arctic Circle, which is at a latitude of approximately 66.5 degrees north. If you want to trace this latitude on a map, then the easiest place to begin is Iceland since the Arctic Circle lies just north of the main island and continues thereafter for some 16,000 kilometres (9,900 miles) to encompass vast areas of Canada, Russia, and Greenland (Figure 2(a)). This line is widely used to identify proximity to the Arctic as a circumpolar region. Many biogeographers would argue that the Arctic tree line, beyond which everything to the north is a landscape characterized by shrubs, dwarf trees, and lichen, is a robust indicator of Arctic-ness. Finally, it is not uncommon to read that the Arctic should be defined by wherever the average daily summer temperature does not exceed 10°C. The last two definitions describe a shrinking Arctic.

The word Arctic comes from the Greek *arktos*, meaning bear, and the identification and naming of the Ursa Major (Great Bear) constellation, which is permanently visible in the Northern Hemisphere sky. In Greek mythology, this constellation is the work of Hera, the wife of Zeus, who in a jealous rage turned the beautiful nymph Callisto into a bear. The North Star, or Polaris, sits immediately above the Arctic pole, with all the night sky revolving around it.

2(a). The Arctic as a geopolitical circumpolar region.

The Arctic Circle marks the most southerly point which experiences to twenty-four hours of daylight during the summer solstice in June and up to twenty-four hours of darkness during the winter solstice in December. Once you get beyond late December, daylight in the Arctic begins to lengthen and by late January, darkness may not make its presence felt until early afternoon. As many residents of the Arctic will affirm, spring and summer are special seasons after the long spell of winter cold and

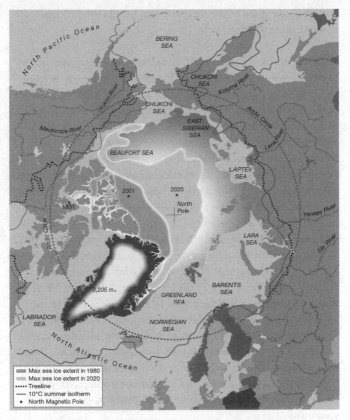

2(b). The Arctic defined by physical and biogeographical parameters.

darkness. Even in the midst of the polar winter, however, moonlight reflects off the ice and snow, so that it is perfectly possible for animals and humans to travel. During daylight hours, even in late autumn and early spring, one's eyes need to be protected from 'snow blindness'. While Inuit developed ingenious snow-goggles made out of animal bone or woolly mammoth ivory, animals such as Arctic fox, polar bears, and reindeer have developed protective mechanisms that enable their eyes both to withstand the glare of UV light and to adjust to the darkness of the long polar night.

If we take the Arctic Circle as the starting point of the Arctic, the region covers about 5 per cent of the Earth's surface—an area of some 20 million km² or 7.7 million square miles. Around two-thirds of the Arctic is ocean, and polar-centred maps are much more helpful than traditional views in conveying the geography and scale of Arctic islands, seas, and ice (Figure 2(b)). The Arctic Ocean is about the same size as the southern polar continent Antarctica. Greenland is the world's largest island, and Baffin and Ellesmere Islands, to the north of mainland Canada, are the fifth and tenth largest islands respectively.

The top of the world reveals the constantly shifting magnetic field of the Earth. There is a *magnetic* as well as *geographical* North Pole. The latter is fixed at 90° N while the former wanders in accord with the behaviour of the Earth's magnetic field. When magnetic north was first identified by James Clark Ross in the 1830s, it was located in the Nunavut territory of Arctic Canada. Since the 1990s it has been moving towards Russia rather rapidly at a rate of 50–60 km each year. At the geographical poles, there are no time zones. Although the poles both north and south help to define global time zones, they have not been allocated one. So, when you reach the geographical North Pole, you can choose your own.

Arctic states

The political geography of the Arctic is dominated by the eight Arctic states: Canada, Denmark/Greenland, Finland, Iceland, Norway, Russia, Sweden, and the United States/Alaska. Russia and Canada are by far the largest. Around half of the circumpolar Arctic is Russian territory, and about 40 per cent of Canada is considered Arctic. Iceland is by far the smallest in terms of both land area and population—most of the country in fact lies *outside* the Arctic Circle. Iceland's total population of around 350,000 is comparable to that of the largest Arctic cities such as Murmansk. While the Arctic is commonly defined as everything north of the Arctic Circle, countries such as Canada also define the 'north' as everything above the 60th parallel. The north and Arctic can and do get used interchangeably, which makes it potentially confusing when one comes to think about national northern territories and their populations. Sweden's northernmost province, Norrland, includes cities and territories along and beyond the 60th parallel.

The governance of Arctic peoples and territories varies markedly across the region. The autonomous status of indigenous peoples is arguably least developed in Russia, partly because the Russian Arctic is considered integral to national resource and security planning. Successive leaders of the Soviet Union and now the Russian Federation recognized that the immense resource wealth of the region needed to be developed, exploited, and protected. 'Resource cities' were established all over the Russian north for the sole purpose of extracting minerals and fossil fuels. Vast areas were simply declared out of bounds and off limits to Russians, indigenous and non-indigenous. The contrast with other parts of the Arctic is stark. In the Nordic region, there are indigenous parliaments in Norway, Finland, and Sweden. The government of Greenland is composed of elected representatives who enjoy a high level of autonomy from the Danish government in Copenhagen. In Alaska and northern Canada, native communities hold substantial resource and land rights and are politically active in government and wider society.

Arctic states and indigenous peoples collaborate with one another internationally in the circumpolar governance of the region. The most notable achievement was the establishment of the Arctic Council via the 1996 Ottawa Convention. As an intergovernmental forum, it created an opportunity for the eight Arctic states and six permanent participants (indigenous peoples' organizations) to work together in areas such as sustainable development and environmental protection. Regionally, the Arctic is a mosaic of governance involving bodies such as the Nordic Council, Northern Forum, Barents Euro-Arctic Cooperation, and European Union. The Arctic also attracts considerable global interest and there are many non-Arctic states such as China and India as well as intergovernmental, inter-parliamentary, and non-governmental organizations (NGOs) who act as observers to the Arctic Council.

People and the Arctic

Some four million people live north of the Arctic Circle but only about 10 per cent are indigenous. Settler communities dominate all of the Arctic, with the exception of Greenland, which is around 90 per cent indigenous. Beyond Greenland, the concentration of indigenous peoples is most notable in the Canadian territory of Nunavut and parts of Northwest Territories. Two million people live in Arctic Russia and the majority are non-indigenous. They live in cities and towns, many of which owe their origins to Soviet-era industrialization and resource development planning. The older imperial city of Murmansk, for example, with a population of around 300,000 people, is a little larger than the city of Anchorage in Alaska (290,000).

The Arctic is a collection of homelands. The indigenous peoples in Alaska, Canada, and Greenland are known as the Inuit, and the indigenous language of that circumpolar community is Inuktitut. Indigenous peoples have lived in the Arctic for around 30,000 years with some of the earliest evidence of human presence found in north-east Siberia. These Upper Palaeolithic people hunted woolly mammoth, woolly rhinoceros, and reindeer. The archaeological record shows that the earliest indigenous cultures of the last ice age were remarkably creative in their production of tools, clothing, and carvings.

Today Russia's indigenous Arctic peoples are estimated to number around 270,000 out of the total population of 2 million. Over 40 indigenous groups live in the Russian Federation and these include the Nenets, Mansi, and Khanty in north-west Russia; the Evenki, Even, and Sakha from Siberia; and the Koryak, Chukchi, and Siberian Yupik who hail from the Russian Far East. In North America, indigenous Arctic peoples are found in Alaska and the Canadian province and territories of Yukon, Northwest Territories, and Nunavut respectively. In Alaska there are Aleuts, Alutiit, Yupiit, and Iñupiat peoples and in Canada there are Athabascan peoples as well as Gwich'in communities that live across the Alaska and Canada borders. Inuit are a circumpolar people living across Alaska, Canada, Greenland, and Chukotka in Russia, and Sámi live in Fenno-Scandinavia, namely Norway, Sweden, Finland, and north-west Russia.

The indigenous peoples of the Arctic have interacted via trade, inter-marriage, and even conflict for thousands of years. In more recent times, these communities have struggled to improve their civil, cultural, political, and resource rights in the Arctic states in which they reside. For centuries, indigenous peoples have been discriminated against and told to stop speaking their own languages and wearing their traditional clothing. They have also been subject to dispossession, with repeated examples of indigenous peoples being forced off their lands and banished from coastal waters. In recent decades, Arctic peoples have secured greater autonomy over traditional lands and their resources by organizing themselves politically in national and international bodies. The Sámi Council, the Inuit Circumpolar Council, and the International Aleut Association are three examples of international bodies that work to promote the interests and wishes of the Arctic's indigenous peoples. The indigenous peoples of the Arctic also live and work in southern cities such as Toronto, Oslo, Seattle, and Moscow. While many indigenous community members still engage in traditional subsistence hunting and reindeer herding, others study at university and work as lawyers, accountants, or serve as elected political officials. There is no one indigenous experience. Many engage with Facebook and

Instagram while herding, hunting, and harvesting fish, reindeer, whales, seals, and walrus.

Estimating the numbers of indigenous peoples in the Arctic is not straightforward. Electoral registers and census data have all been used with varying success to identify the number and distribution of people. Many demographers would not focus on ancestry alone because there are many established Arctic communities with mixed heritage. It is quite common in Greenland, for example, to meet Greenlanders with a Danish father and a Greenlandic mother or vice versa. In Russia, indigenous groups are not recognized as such if they exceed 50,000 in number. By defining groups as indigenous on the basis of size, such as the so-called 'small peoples of the North, Siberia and Far East', larger groups such as Yakuts, Komi, and Komi-Permyak, who would be considered indigenous elsewhere in the Arctic, are excluded.

Understanding the Arctic

The Arctic is a physically diverse, socially and ecologically rich, and politically complex region. Even the geographical definition of the Arctic can become confused with a more generic definition of the 'North'. What we can note with confidence is that there are two distinct Arctics. The first conception recognizes the Arctic as a living and working homeland. It is an Arctic with a rich tapestry of human history and a place where intimate relationships between people, plants, animals, and landscapes have evolved. It is an Arctic that is broadly cold and dark in the winter, but across which the climate varies greatly, from the very cold environments of Alaska, Greenland, and Siberia to the milder and wetter Nordic Arctic. In winter, it is not uncommon for Alaska to experience –30 to –40°C, and in the Russian north the air temperature can drop even lower. Thanks to the Gulf Stream, Iceland and the northern parts of Norway are the mildest elements of the Arctic. In the short but intense summer, the Arctic is home to many migratory species including birds and whales. Over millennia, people and communities have become skilled at harvesting marine mammals, fish, wild plants, musk oxen, and caribou. Depending on where you are in the Arctic, the look and feel of Arctic communities and ecologies can be very different.

The second is the Arctic of imagination. The land, ice, and water of the far north have long been the subject of fable, myth, and speculation. In the Greenlandic summer, long-established coastal communities travelled inland (Nunap Timaa) in pursuit of caribou and fish. But they did so with trepidation because the interior of Greenland was believed to be filled with ghosts, human outcasts, monsters, and giant peoples. From the Ancient Greeks and Vikings to 19th-century English romantic writers and 21st-century filmmakers, the Arctic has been imagined to be a source of romance, danger, and intrigue. Individuals and groups have travelled there, extracted saleable goods such as walrus ivory and oil, and returned with stories and images of monstrous sea ice, bitterly cold weather, and spectacular skies, often hung with the strange glows and shifting curtains of the *aurora borealis* or northern lights. Indigenous mythologies tell their own stories about the long polar night, including a Finnish Sámi tale about a fire fox who ran so quickly that his tail caused sparks to fly into the dark skies. Nowadays our images and stories about the Arctic have become far more pessimistic and warn about the ongoing impacts of planetary warming on sea ice, permafrost, and the prevailing climate system.

Two fundamentals are changing markedly in today's Arctic. First, the lived experience of Arctic communities is being turned upside down. For thousands of years, indigenous communities have organized their lives around a cold climate, with a short but intense summer season. This way of life is altering because sea ice is disappearing, frozen ground is thawing, and wildfires are burning northern forests and peatlands. Arctic waters have not been spared alteration, and as the relationship between sea ice and ocean shifts, it carries troubling implications for the availability of marine mammals and fish. Second, there is growing global interest in the Arctic which goes beyond romantic storytelling and indigenous mythologies and eyes its rich resources and strategic value. Arctic communities and states have had to accommodate outside powers such as China, India, and the European Union, all eager to secure their economic and strategic interests.

Looking ahead, we detect three major drivers of change. First and foremost is ongoing environmental change. What lies in store for the key building blocks of the Arctic landscape—permafrost, sea ice, and glaciers—within a rapidly warming climate? How will Arctic ecosystems fare on land and in the seas? Is the Arctic carbon budget shifting from sink to source? How will the indigenous peoples of the Arctic adapt to these profound changes?

The second is resource potential. The Arctic is rich in natural resources including fisheries, timber, minerals, rare earth elements, oil, and gas. But economic development is highly uneven. Resource extraction is a major source of employment and wealth creation for northern communities, Arctic states, and global investors, including sovereign wealth funds and third parties such as China. There is considerable nervousness about the exploitation of Arctic resources. In July 2020, China's Shandong Gold Mining Company considered buying a Canadian gold mine in Northwest Territories. Concerns have been raised that China's economic and strategic presence needs to be restrained rather than encouraged.

Finally, as the Arctic attracts ever more global attention, it brings to the fore complex issues of governance. Russia and the United States are boosting investment in polar infrastructure and sending their naval forces into Arctic waters. New shipping lanes are opening up for global trade. While China is a self-declared 'near-Arctic state' with a growing portfolio of trade and political interests, the Arctic's indigenous peoples are demanding that their legal rights and political aspirations be taken seriously. As the Arctic remains integral to environmental campaigning and civil society activism, we should expect no let-up in demands for the Arctic to be better protected. If the world wishes to 'leave resources in the ground' then it needs to be recognized that many Arctic communities are highly entangled in resource-dominated livelihoods. What lies in store for the Arctic and its landscapes, ecosystems, and people?

Conflict is not inevitable in the Arctic, but neither is cooperation and goodwill. Today's Arctic is tomorrow's world. As Arctic landscapes and societies are reshaped, the adaptation and governance issues that emerge will offer all of us a glimpse of the many hard choices that will follow.

Chapter 2
The physical environment

Geological and tectonic setting

Millions of years of plate tectonic movements have arranged land and ocean in opposing ways in Earth's polar regions. While the Antarctic continent is surrounded by the Southern Ocean, the Arctic is its polar opposite: an ocean semi-enclosed by land. About one-third of the area within the Arctic Circle is land, another is shallow shelf seas, and the central third is deeper ocean basin. Some of Earth's most ancient rocks—over 4 billion years old—are found in northern Canada and Greenland. The rocks of the Arctic bear witness to key periods in Earth history. Svalbard, for example, contains some of the most detailed geological evidence for the Snowball Earth events of the Cryogenian Period (720–635 million years ago) that saw worldwide glaciation. Much later, fossil-rich sandstones and marine shales accumulated in large sedimentary basins from the Carboniferous Period (359–299 Ma) onwards, producing the oil- and gas-trapping rocks that have attracted the petroleum industry to the Arctic. Great thicknesses of basalt, known as the Siberian Traps, were laid down during a colossal phase of volcanism that has been linked to the end-Permian mass extinction some 252 million years ago. These lavas cover about 5 million km^2 of the northern Siberian shield—about one-third of the Russian landmass. Much later, in the balmy climates of the Cretaceous Period (145–66 Ma), dinosaurs were part of humid forest and swamp ecosystems inside the Arctic Circle. The northern parts of our world haven't always been frozen.

The Arctic Ocean basin is bisected by the Lomonosov Ridge: a steep-sided submarine mountain range more than 1,800 km long that runs under the North Pole (Figure 3). The ridge protrudes up to 3,000 m from the ocean floor to separate the Canadian (western) and Eurasian (eastern) marine basins—it was named for the 18th-century Russian polymath Mikhail Lomonosov (1711–65), who put forward the first theory of iceberg formation in 1760. A Russian flag was planted on this ridge in the summer of 2007 at a depth of 4,200 m beneath the pole. A narrow belt of seismic activity runs parallel to the Lomonosov Ridge tracing the divergent tectonic boundary where the North American and Eurasian plates are prised apart (Figure 3). This is the Gakkel Ridge where, thousands of metres below the sea ice, super-heated basalt lavas forge new oceanic crust. Gakkel is an ultraslow spreading ridge with a mean rate of about 1 cm per year: the slowest of any mid-ocean ridge so far observed on Earth. Submarine volcanoes and hydrothermal vents have been recorded along this polar plate junction. The Gakkel ridge is the most active source of earthquakes in the Arctic: geophysicists have stowed seismometers on ice floes to record the seafloor tremors.

When did the Arctic get cold?

The marine sediment record indicates that some winter sea ice may have first appeared during the Eocene Epoch around 47 million years ago, following the onset of global cooling after a period of unusual greenhouse warming known as the Palaeocene–Eocene Thermal Maximum (PETM). Sea ice is frozen seawater that floats on the ocean surface. A combination of factors set the Earth on a course of long-term cooling after the PETM. These included the isolation of the Antarctic continent during the subsequent Oligocene Epoch (33.9 to 23 Ma) and the rise of the Himalayas that resulted in huge amounts of weathering under a monsoon climate drawing carbon dioxide from the atmosphere and weakening Earth's greenhouse effect. As the extent of ice and snow steadily increased in the polar regions, feedbacks intensified the cooling trend, and the cryosphere became a major player in the Earth system. While large-scale glaciation began in the Oligocene in Antarctica, it developed rather later in the Arctic.

The offshore sediment record contains evidence of iceberg calving from Greenland in the Miocene (23 to 5.3 Ma), around 18 million years ago, thus indicating the presence of glaciers at sea level and extensive glaciation at this time. By 13 to 14 million years ago, there are indications that multiyear sea ice had developed in parts of the Arctic Ocean. This is sea ice that has survived at least one melt season. It was much later, however, following the marked global cooling that set the stage for the Quaternary ice ages (2.58 million years to the present Holocene interglacial), that extensive and perennial sea ice cover became established in the Arctic on a scale comparable to that of the historical era. *Kryos* is Greek for cold and the cryosphere includes sea ice, glaciers, snow, permafrost, lake and river ice, hail, and frost—frozen H_2O in all its myriad forms. Today's Arctic displays remarkable variety in its frozen landscapes—Russian scientists have coined the term *cryodiversity* to capture this richness.

Arctic climate

The Arctic is cold today because the polar latitudes receive much less solar energy per unit area than the rest of the Earth's surface and snow and ice reflect a good deal of this energy back into space. The Arctic sun is always low in the sky and solar rays travel a greater distance to reach the poles. All of this means the polar latitudes experience a pronounced solar radiation deficit. Because of the tilt of the Earth, this energy deficit has a very strong seasonal component. During the boreal winter, when the Northern Hemisphere is angled away from the sun, sunlight cannot reach the far north and the High Arctic experiences the long polar night with several months of darkness. The polar night increases in length by about six days for each degree of latitude north of the Arctic Circle. Thus, it lasts for about one month at 68° N and close to six months at the pole.

The oceans and the atmosphere attempt to even out the planetary energy imbalance by moving heat towards the poles. Atmospheric circulation is a key control on Arctic weather and climate, especially in winter when solar warming is absent. During the long polar night, atmospheric circulation is the dominant supplier of warmth to the region (~95 per cent) with the rest from ocean currents. Depending on place and season, precipitation and temperature can show great variability: contrast, for example, the very cold and dry continental climate of much of Siberia with the maritime conditions of Arctic Norway where the coast receives warmth from the Gulf Stream. More generally, precipitation is low (150 to 250 mm) across the Arctic and the most northerly landscapes are polar desert. Mean summer air temperatures sit above freezing in all parts of the Arctic apart from the core zone of multiyear sea ice around the pole. Winter temperatures below −60°C have been recorded on the high interior of the Greenland ice sheet: these are the coldest anywhere in the Arctic.

As sunlight disappears in late autumn, the Arctic atmosphere becomes extremely cold while the equator remains warm. In response to this thermal contrast, a circumpolar vortex of powerful winds and extremely cold air forms in the upper troposphere and stratosphere, more than 10 km above the Earth's surface. It marks the latitudinal boundary of cold polar air and warmer subtropical air masses. It is found around both poles, turning counter-clockwise, and is a key feature of high-latitude meteorology. It is strongest in midwinter when the thermal contrast between equator and pole is at its maximum. This vortex of low pressure helps keep cold air locked in the far north.

The Arctic vortex is vulnerable to disturbance from large-scale weather systems that are influenced by mountain ranges or warm ocean currents. Sections of the vortex can be pulled away so that the meanders become elongated; this can funnel Arctic air into the mid-latitudes as well as allowing warm air into the far north. The term 'bomb cyclone' has been used in North America to describe the resulting extreme drop in atmospheric pressure and intense blizzard conditions associated with a cold Arctic air mass colliding with a warmer air mass. Conversely, incursions of warm air can lead to winter heatwaves in the far north. Each spring, the vortex breaks down when sunlight warms the polar atmosphere and lessens the temperature difference between equator and pole.

The ocean at the top of the world

The Arctic Ocean is the smallest and shallowest of the world's oceans. It covers an area of more than 14 million km^2 within a broadly circular basin centred on the Geographic North Pole (Figure 3). It is semi-enclosed by the northern extremities of Europe, Asia, and North America (including Greenland). Together, the shorelines of these landmasses account for some 45,000 km of coastline including bedrock cliffs, eroding permafrost bluffs, deep fjords, calving glaciers, shingle beaches, river deltas, and extensive wetlands. This is where the treeless northern tundra meets the sea.

Almost half of the Arctic Ocean is underlain by continental shelf (Figure 3). The marginal shelf seas—of which the Barents, Kara, Laptev, East Siberian, and Chukchi are the largest—are a highly distinctive feature of the Arctic marine environment. The physical, biological, and chemical processes operating on these shallow water shelves influence the oceanography of the entire region. These shelves were dry land for much of the Quaternary ice age and many of today's Arctic islands were connected to the continents. At those times when global sea level was much lower than today, the shelves extended parts of the Arctic tundra ecosystem by up to 10° of

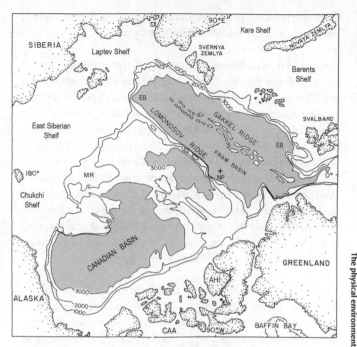

3. **The Arctic Ocean basin showing the shallow shelf seas and submarine ridges (MR = Mendeleyev Ridge; EB = Eurasian Basin; CAA = Canadian Arctic Archipelago; AHI = Axel Heiberg Island; NP = North Pole).**

latitude so that woolly mammoths, musk oxen, and reindeer foraged across a landscape that is now submerged. During glacial periods, the far north-east of Siberia was connected to North America via the Bering Land Bridge to Alaska. These shelves have accumulated great thicknesses of sediment; they form an important depositional sink for river sediment, dead marine biota, and ice rafted debris. The shelf deposits store large quantities of methane hydrates that formed during the Pleistocene from the breakdown of organic matter. Methane hydrate is an ice-like substance formed when methane (CH_4) and water combine at low temperature under moderate pressure. In the nearshore shallows, sea floor sediments are often scarred by the keels of drifting ice floes.

Because of its remoteness, sea ice cover, extreme weather, and months of darkness, the Central Arctic Ocean is perhaps the world's least well-studied marine environment. There is still much to learn about its currents and the deepest basins beneath the perennial sea ice. We have better topographic data for parts of Mars than we do for the contours of the Arctic abyss. Rates of sediment deposition on the floor of the Central Arctic Ocean are low, typically less than a few centimetres per thousand years. We know very little of the oceanography, ecology, and recent geological history of the deepest parts of this ocean. Nevertheless, some of these knowledge gaps are being filled as digital mapping of the Arctic Ocean floor progresses. We will no doubt discover much more as the sea ice continues its steady decline.

The Arctic is the least saline of the world's oceans, due to limited evaporation from its cool surface and huge seasonal inputs of freshwater from the great Arctic rivers and glacier melt. Three of the largest rivers on Earth, the Ob, Yenisey, and Lena, drain a vast area of the Eurasian landmass. The Yukon and Mackenzie are the largest rivers in North America that empty into the Arctic (Table 1). Runoff into the Arctic Ocean from the Eurasian rivers is three to four times greater than that from North America.

Under 'normal' conditions, the Arctic Ocean has a distinctive vertical structure—the surface waters (the upper 50 m or so) are colder and fresher than the deep waters, and this helps sea ice to form. These upper and lower water masses are separated by a steep salinity gradient known as a halocline—the ocean becomes saltier and therefore denser with depth. The halocline plays a critical role in maintaining stratification and limiting heat transfer upwards to the surface. The importance of a stable stratification becomes clear if we consider that the salty Atlantic waters that

Table 1. The great Arctic rivers ranked by water discharge

River	Drainage basin area (million km²)	Mean annual water discharge (km³/year) and global rank	Permafrost area (%)	Receiving waters
Yenisey	2.5	673 (5th)	88	Kara Sea
Lena	2.4	581 (6th)	77	Laptev Sea
Ob	2.95	427 (13th)	26	Kara Sea
Mackenzie	1.8	316 (18th)	82	Beaufort Sea
Yukon	0.83	208 (22nd)	>95	Bering Sea
Kolyma	0.65	136 (29th)	100	East Siberian Sea

penetrate the Arctic at depth contain enough heat to melt all the sea ice within just a few years.

There are three main connections between the Arctic and the global ocean: the Bering Strait connects to the Pacific, and marine gateways to the west and east of Greenland connect to the Atlantic (Figure 1). One of the latter is the Fram Strait—a very deep channel that lies between the north-east coast of Greenland and the Svalbard archipelago. It is the only deep connection between the Arctic and the global ocean. Two currents flow side by side: the East Greenland Current moves cold waters and sea ice out of the Arctic whilst the West Spitsbergen Current moves warm water northwards from the Atlantic. At its northern end, the Fram Strait includes the Molloy Deep which, at some 5,550 m below sea level, is the deepest point in the Arctic Ocean. The Chukchi and Barents shelves are strongly influenced by inflowing waters from the Pacific and Atlantic, respectively, whilst the Siberian shelves come under the influence of sediment and water supplied by the big Russian rivers—most notably the Ob, Yenisey, and Lena (Table 1).

Arctic sea ice

Sea ice is a fundamental component of the Arctic environment. It can be fixed to the coastline as a stable platform (landfast ice) or drift with ocean currents in deeper water. Beyond the core of multiyear ice, the growth and retreat of Arctic sea ice follows a characteristic seasonal cycle: it forms throughout the boreal winter reaching peak coverage in early March followed by steady summer melting to the annual minimum extent in September (Figure 4(a)). Before the advent of routine pan-Arctic monitoring by satellite in 1979, data on sea ice coverage was patchy and involved multiple sources including logbooks from whaling ships, newspaper reports, and observations from aircraft, as well as more systematic regional surveys carried out by Arctic nations.

Today the National Snow and Ice Data Center (NSIDC) based at the University of Colorado and funded by the US government is the key body collecting and archiving sea ice data from both hemispheres. The NSIDC supports research into all aspects of the cryosphere and has developed particular expertise in satellite monitoring of snow and ice. Its records show that, historically, Arctic sea ice extent has been about 14 to 16 million km^2 at the end of winter before falling to about half that area by early September. There is currently much concern about the steep decline in Arctic sea ice observed over the last four decades. In 1980 the September Arctic sea ice *minimum* was 7.67 million km^2. In September 2020 it fell to 3.74 million km^2—the second lowest on record (Figure 4(b)). Relative to the average coverage observed during the period 1981 to 2010, the September sea ice minimum is now falling by about 13 per cent each decade. This trajectory brings sea-ice-free summers to the Arctic Ocean by the middle of the present century. The September value for 2012 (3.57

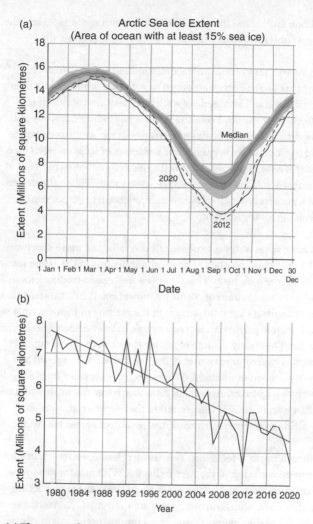

4. (a) The seasonal march of sea ice development in the Arctic. (b) The steady decline in mean monthly Arctic sea ice extent in September during the satellite era (post 1979).

million km²) is less than half the sea ice coverage of 1980 and the lowest observed to date in the satellite record (Figure 4(b)).

There are many ramifications of dwindling sea ice cover. The amount of solar energy absorbed or reflected is a key control on Arctic climate. The albedo of a surface measures how well it reflects solar energy. From the Latin *albus* (white), this physical property is of fundamental importance in the Arctic because ice and fresh snow are very effective reflectors of solar energy. Open water is a very poor reflector, reflecting only about 10 per cent of shortwave solar radiation. Sea ice has a much higher albedo, and sea ice covered with fresh snow can reflect up to 90 per cent of incoming solar radiation.

As more open water is exposed, albedo falls, and more shortwave solar energy is absorbed. The resultant heating of the ocean leads to further sea ice melt. This is the sea ice–albedo feedback loop and a key component of Arctic amplification. While this feedback loop has always been important in the melting of thin seasonal sea ice, it is now playing an increasing role in the demise of *perennial* sea ice. The steady decline of sea ice extent in the polar summer leaves ever greater expanses of dark ocean surface exposed to twenty-four-hour sunlight. Where sea ice is replaced by ocean, the energy budget is transformed. As sea ice cover diminishes and the ocean warms, heat is transferred to the overlying atmosphere resulting in warmer surface air temperatures across the Arctic.

Other important feedbacks can also accelerate sea ice decline and promote warming of the Arctic climate. With greater expanses of open water, larger waves can form to break the ice pack into smaller floes, which further speeds up their melt. As Arctic storms generate higher energy waves, more intense ocean mixing draws warmer waters to the surface as stratification breaks down. These waters not only melt sea ice, but they also delay the formation of new sea ice in the winter and extend the period of ocean surface warming. Sea ice is under attack from all directions.

Sea ice can be classified by age and thickness, but direct observations of the latter are limited. Before satellite monitoring, Soviet and US submarines gathered sonar data on Arctic sea ice thickness, but most were hidden in secret military files. Declassified records indicate that mean sea ice thickness was about 5–7 m in the 1960s and 1970s. Today it is about 2–3 m. In the summer, the upper surface of multiyear ice melts. Then, in winter, the ice pack thickens again when new ice forms at its base. The Central Arctic Ocean (CAO) covers an area of about 2.8 million km^2 and has been dominated, historically, by the thickest multiyear ice. In the 1980s, just 1 per cent of the CAO was open water at the seasonal sea ice minimum in September. Between 2010 and 2017, however, the extent of open water in the CAO had risen to more than 20 per cent. The oldest and thickest sea ice is also in decline. Today this sea ice is found in a belt stretching for some 2,000 km from the north coast of Greenland to the western end of the Canadian Arctic Archipelago. This region has been called the Last Ice Area—its status is critical for the maintenance and health of sea-ice-dependent ecosystems.

The fourteen years before 2020 saw the fourteen lowest sea ice extents in the satellite record. Arctic sea ice is now also younger, thinner, and more mobile than at any time in the satellite era. Perhaps the most direct impact on Arctic communities is the continued long-term decline (since the 1970s) in coastal landfast sea ice. This ice protects coasts from wave erosion and provides a vital platform for hunting expeditions and travelling. In the Kara, Laptev, and East Siberian seas, landfast ice can extend for hundreds of kilometres from the coast and in shallow waters it can even freeze to the seabed. The multiple channels of the Canadian Arctic Archipelago form a network of landfast ice connecting the entire region. On open coasts, landfast ice can be buffeted and deformed by drifting pack ice to form prominent bulges known as stamukhi.

Sea ice is a fundamental building block of the Arctic ecosystem. A specialist group of Arctic mammals are uniquely adapted to on sea ice and this includes the polar bear (*Ursus maritimus*), walrus (*Odobenus rosmarus*), Arctic ringed seal (*Pusa hispida hispida*), and several other species of seal. All aspects of their feeding habits, reproduction strategies, and resting revolve around the presence and quality of the sea ice edge. Light transmission increases as sea ice thins and in the presence of melt ponds. This is a key control at the bottom of the food chain since algae that grows on the underside of sea ice does so by absorbing sunlight.

Further out to sea, beyond the ice anchored to the coast, sea ice is always in motion. The forces of wind and ocean currents generate enormous pressures that can deform the ice surface and create substantial relief. Like tectonic plates in miniature, where ice floes collide under pressure, linear ridges emerge where ruptured slabs are forced both upwards and downwards. The distance from ridge top to keel can be tens of metres on the largest pressure ridges. These trap wind-blown snow and can survive for several years.

Large areas of open water where one would normally expect thick sea ice are called polynyas. These are often found in the shallow coastal zone where strong offshore winds push the sea ice out into deeper water. These polynyas function as important sea ice nurseries. As salt is released during sea ice formation, the local density of seawater is increased, and these shelf seas can be an important generator of salty deep water for the Arctic Ocean. Sea ice forms throughout the long Arctic winter in all the shallow seas off northern Russia where air temperatures can drop below −40°C. This Arctic shelf is a key source area for new sea ice, where powerful winds from the Siberian landmass push ice floes northwards into deeper, open water. This Russian ice is then shunted across the central Arctic by the wind-powered transpolar drift and eventually into the Fram Strait, where it melts. This journey takes two to three years. Historically, the Russian shelf has been an important provider of young sea ice to the central Arctic, but satellite records show that this supply is now in decline. Siberian sea ice hit record lows in 2020. Two decades ago about half of the sea ice that formed in the shallow seas off Russia made it all the way to the Fram Strait. Today this figure is about 20 per cent. The Fram Strait is the most northerly part of the global ocean that is free of sea ice throughout the year.

Permafrost

Ground that remains frozen at or below zero Celsius for at least two years may be called permafrost. It is one of the defining characteristics of the Arctic landscape. This iron-hard ground covers almost half of Russia and Canada, accounting for about one-quarter (23 million km^2) of the entire land surface in the Northern Hemisphere (Figure 5). Whether in soil, sediment, or bedrock, permafrost is classified according to its areal extent as continuous (90–100 per cent), discontinuous (50–90 per cent), sporadic (10–50 per cent), or isolated (0–10 per cent). The importance of continuous land and sub-sea permafrost in the high northern latitudes is clear from the map in Figure 5. Note how a large part of the European Arctic is characterized by sporadic permafrost. Scandinavia and Iceland have relatively little permafrost because of their proximity to warm ocean currents. In upland landscapes to the south of the continuous permafrost zone, frozen ground can be restricted to north-facing slopes. The map does not show permafrost that may exist beneath the Greenland ice sheet.

Permafrost can vary in thickness from less than a metre to many hundreds of metres: in parts of Yakutia in eastern Siberia, the permafrost penetrates to more than 1,500 m below ground. This very deep permafrost is often relict—it formed during colder climates of the ice age past and could not form to such depths under present-day conditions. The permafrost beneath the shallow marginal seas of the Arctic Ocean is also relict and formed during the lowered sea levels of the Pleistocene glacials when these areas were dry land. The upper surface of this permafrost normally lies a few metres below the seabed, but the extent of its vulnerability to ocean warming is not fully understood.

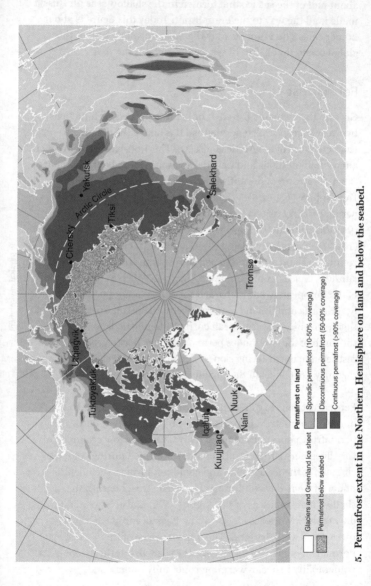

Permafrost on land

Sporadic permafrost (10–50% coverage)

Discontinuous permafrost (50–90% coverage)

Continuous permafrost (>90% coverage)

Glaciers and Greenland ice sheet

Permafrost below seabed

5. Permafrost extent in the Northern Hemisphere on land and below the seabed.

28

Since the last glacial period, most of the Arctic landscape has been influenced by cold, non-glacial conditions and can therefore be termed a periglacial environment where the majority of ice is actually below ground. Periglacial landscapes are influenced by intense frost shattering—often in the presence of permafrost—and this can result in a distinctive assemblage of landforms including patterned ground, such as tundra polygons created by thermal contraction cracking in winter, extensive block fields, and thick screes resulting from freezing and thawing of bare rock surfaces. Where frost action sorts the debris, stone stripes and stone circles decorate the landscape (Figure 6(a)). Ground ice occupies pore spaces, veins, and cavities in soils and bedrock—it can form massive ice wedges that penetrate several metres below ground and distinctive ice-cored rounded hills known as pingos (Figure 6(b)). Hundreds of pingos are found in the lower reaches of the Mackenzie River in Northwest Territories where the largest reach 70 m above the surrounding floodplains. Wind action is also important as an erosive agent, a transporter of dust and sand, and a creator of dunes.

The permafrost of Siberia is generally thicker and more extensive than that in Arctic North America—this is partly explained by their contrasting glacial histories. Most of eastern Siberia remained free of glacial ice during the Pleistocene. It was certainly cold enough for glacier development, but glaciers will not form without sufficient snowfall, however intense the winters. Much of Canada was covered by the enormous Laurentide and Cordilleran ice sheets; these were nourished by abundant supplies of moisture from maritime air masses. Glacial processes repeatedly scoured the Canadian landscape, often down to bedrock, and created thousands of lakes following deglaciation. Across much of Siberia, however, beyond the high mountains, snowfall was too low and did not survive the summer. For a good deal of the Pleistocene (2.58 million years to 11,700 years ago), vast regions were exposed to bone-chilling winds and very cold air temperatures with short summers. This dry and frigid climate allowed continuous

The Arctic

6. Periglacial landscapes in the permafrost. (a) Frost-shattered rock debris arranged in stone circles around collapsed ice lenses in Svalbard. (b) A collapsed pingo in the Mackenzie Delta of Arctic Canada surrounded by wetland. Ground ice beneath the pingo is insulated by a thick cover of peat.

permafrost to develop to great depths in Siberia, especially where the deposition of sediment by rivers and silt-bearing winds kept adding to the thickness of the frozen ground.

If the near-surface seasonally thaws and freezes, it is known as the active layer. It is not part of the permafrost. The active layer can vary in thickness from a few centimetres to a metre or more depending on the intensity of warming and local conditions. It is important because hydrological, ecological, and biogeochemical processes are concentrated in this layer. The thickness of the active layer generally increases as latitude decreases but is also influenced by sediment type and vegetation cover. Most of the permafrost actually lies in the boreal forest zone (roughly 50° N–70° N) where trees shield the ground from solar warming and limit the depth of the summer thaw.

Since permafrost is impermeable it exerts a profound influence on the geomorphology and hydrology of the Arctic landscape. During the summer thaw seasonal ponds and waterlogging become key components of Arctic ecosystems. Wetlands come alive across the Arctic, but the thaw can also lead to ground instability, water erosion, and mass movements. The downslope creep of active layer debris is called solifluction, and it can lead to the formation of distinctive lobes on hillslopes.

Canadian Earth scientists have identified some very old permafrost in the form of relict ground ice preserved within the discontinuous permafrost of central Yukon. They were able to establish a minimum age for a large ice wedge complex because it was capped by volcanic ash known as the Gold Rush tephra. By dating the ash to 740,000 ± 60,000 years ago, they have documented the oldest ice known in North America. The ice wedge may have formed in the Lower Pleistocene. This study is important because it demonstrates that permafrost has been a component of the North American cryosphere for a very long time. It also shows the resilience of this permafrost to past interglacials that were longer and warmer than the Holocene. This finding suggests that deeper lying permafrost—more than a few metres below the ground—may be more stable than previously thought.

As the summer thaw now begins earlier, lasts longer, and penetrates deeper into the ground, the Arctic landscape is responding in dramatic ways. The impacts of permafrost thaw are increasingly evident at the coast where the disappearance of landfast sea ice has left thawing cliffs exposed to wave erosion. As blocks of land tumble into the sea, coastlines recede, and coastal waters see increased turbidity. Mercury can be released to the atmosphere and into surface runoff as soil organic matter decays. The degradation of ground-ice-rich permafrost creates *thermokarst*—this is a landscape of shallow lakes and unstable ground prone to slumping and other kinds of mass movement. Lakes form and empty rapidly as local hydrology adjusts to new ground conditions. In south-west Alaska, the townsfolk of Kongiganak stopped burying their dead about ten years ago when thawing of the permafrost turned the local cemetery into a swamp. They now place painted wooden coffins above the sodden ground. A state of emergency was declared in Russia in June 2020 when 21,000 tonnes of diesel oil spilled into the Ambarnaya River near the Arctic city of Norilsk. The catastrophic failure of the storage reservoir was partly attributed to thawing ground. With a temperature rise of 4°C in the Arctic, it has been estimated that the southern permafrost boundary could shift northwards by up to 500 km. Permafrost thaw will be costly across the Arctic as authorities seek to protect energy infrastructure, buildings, roads, and railways from further disruption.

High Arctic glaciers and the Greenland ice sheet

Glaciers of all kinds can be found in the High Arctic from isolated mountain cirques to the vast Greenland ice sheet. The islands of the Barents and Kara seas—including Svalbard, Franz Josef Land, Severnaya Zemlya, and Novaya Zemlya—are all extensively glaciated. Glaciers and ice caps are also present on the large islands of the Canadian Arctic Archipelago (CAA). Glaciers can form at low elevations in the Arctic because they are so far north. Where they emerge from a valley and spread out across a low-relief plain they are called piedmont glaciers. Those on Axel Heiberg Island that spill out across the braided river plain upstream of Surprise Fjord are truly spectacular (Figure 7). Because snowfall is typically low, the glaciers of the High Arctic tend to accumulate mass very slowly in comparison to alpine and maritime glaciers in the mid-latitudes. These polar desert glaciers are especially vulnerable to even modest increases in summer warmth. The slow growing glaciers of the High Arctic are relics of past climate regimes because they would not form under the present-day warming climate. The Barnes Ice Cap on Baffin Island, Nunavut, contains ice that formed around 20,000 years ago during the last ice age. This is Canada's oldest glacial ice—a remnant of the giant Laurentide Ice Sheet.

7. Piedmont glaciers on Axel Heiberg Island.

Most Arctic glaciers are shrinking—they are just not viable in a rapidly warming Arctic where summer melt far outstrips winter accumulation. The glaciers and ice caps of the CAA account for about one-third of the land ice outside Greenland and Antarctica. In the last few decades, ice losses from these glaciers—and from those in Alaska—have been one of the largest contributors to global sea level rise after the ice sheets.

Greenland is still in the grip of an ice age. Its ice sheet is the largest body of glacial ice on Earth outside Antarctica with a volume of some 2.85 million km^3 extending over 1.71 million km^2 (Figure 8). An ice sheet of comparable dimensions to the one we see today first developed about 3.5 million years ago and has been present throughout the Quaternary Period (2.58 million years ago to the present). The ice sheet forms the largest expanse of elevated topography in the Arctic. It is over 3,000 m thick at its highest point. The ice sheet becomes much thinner towards the coast, where mountains support substantial independent ice caps and valley glaciers that are not connected to the main ice mass.

Because many glaciers extend down to sea level in the polar latitudes, they can lose mass by calving icebergs to the ocean. These are tidewater or outlet glaciers. Marine-terminating glaciers are also sensitive to changes in ocean temperature. Where underwater canyons are cut into the continental shelf, warm deep waters can reach glacier fronts. Enhanced submarine melting is thought to have been the key factor forcing the retreat of many of Greenland's outlet glaciers. Jakobshavn Glacier in western Greenland is one of the most prolific outlet glaciers. It is Greenland's fastest glacier and the source of about 10 per cent of Greenland's icebergs. In a typical year over 35 billion tonnes of ice are shed from its calving front. The 2012 *Chasing Ice* documentary included footage of a spectacular calving event when over 7 km^3 of ice broke away from this glacier.

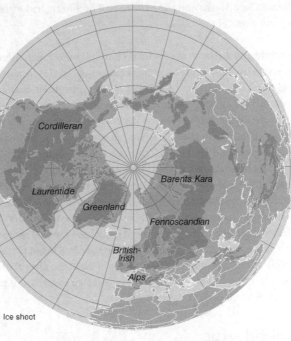

Ice sheet

8. **The maximum extent of glaciation in the Northern Hemisphere in the last glacial period. Much of the Arctic (including large areas that are now below sea level) was covered by glacial ice at this time. Arctic-like conditions also extended into the mid-latitudes.**

At the Last Glacial Maximum (LGM), around 25,000 years ago, the Greenland ice sheet covered all of the land that is currently ice free and extended onto the continental shelf in all directions. This Late Pleistocene ice sheet extended over some 3 million km². Figure 8 shows the extent of glacial ice in the Northern Hemisphere at the LGM. Note how the islands and peninsulas of the Barents and Kara seas—including the Svalbard archipelago and Franz Josef Land—are also covered by a major ice sheet that was contiguous with the Fennoscandian and British-Irish ice sheets. Only Alaska and the eastern half of Siberia had extensive ice-free areas at the LGM.

Satellite and airborne observations have allowed much more robust estimates of mass balance changes in Arctic glaciers including the Greenland Ice Sheet. We monitor their changing area and thickness, map the number and size of supraglacial lakes, measure the velocity of ice streams and the advance and retreat of outlet glaciers in the fjords. We can even map the bedrock topography beneath the ice. The mass balance of the Greenland ice sheet began to deviate in the 1980s from its historical range of variability. Since then, annual ice loss has increased by a factor of six. Ablation is dominated by meltwater runoff from the ice sheet surface and via increases in iceberg calving from tidewater glaciers. Greenland melt has raised global sea level by more than 13.7 mm since 1972 and half of this increase has taken place within the last decade. In the summer of 2019 the Greenland ice sheet lost 600 billion tonnes of ice in just two months and raised sea level by more than 2 mm.

A global Arctic

The Arctic Ocean is integral to the global climate system. Sea ice extent is a fundamental influence on the energy balance of the high northern latitudes and one of the key regulators of global climate. It helps to keep the Arctic cold by reflecting solar energy throughout the polar summer and by limiting heat exchange between the cold atmosphere and the much warmer ocean waters below. Because of the ice–albedo feedback, the dramatic reduction in sea ice extent observed in the satellite era has promoted strong warming across the Arctic. This feedback is so powerful because one of Earth's most reflective surfaces (sea ice with fresh snow) is replaced by one of its least (dark ocean). Mass loss from the Greenland Ice Sheet will continue to be a major contributor to sea level rise this century. If all the Greenland ice melted and flowed to the ocean, it would raise sea level by over 7 m and flood all the coastal cities of the world, including Bangkok, London, New York, and Shanghai.

Arctic waters are important for the formation of cold, dense, ocean bottom waters, especially in the North Atlantic where they help to drive the planetary network of ocean currents known as thermohaline circulation. The Arctic Ocean is the most northerly loop of this global energy flow. It receives waters from both the Pacific and the Atlantic, but all the outflows are to the Atlantic via the deep Fram Strait and through channels to the west of Greenland. The Atlantic Meridional Overturning Circulation (AMOC) is a system of ocean currents that convey heat from the tropics into the North Atlantic. The strength of this circulation relies on the formation of cold, dense, saline water that sinks in the Greenland Sea. The AMOC is weakened when there is a greater flux of lower density freshwater from the Arctic.

The Arctic Ocean is getting warmer and fresher. The largest temperature increases have been observed towards the edge of the continental shelf in the northern Barents and Kara seas. These hotspots have seen the most rapid surface warming and the greatest fall in winter sea ice cover anywhere in the Arctic. In November 2018 the temperature of the bottom waters on the northern Bering shelf rose above 4°C for the first time since records began. Norwegian oceanographers have recently shown that water column stratification is breaking down in the northern Barents Sea and warming is evident at all depths. Sea ice cover in the Arctic is shifting to a new, thinner regime so the ice pack will be both more fractured and mobile. This enables another feedback because increased wave action drives ocean mixing and brings more warm water to the surface. This melts more sea ice and hinders the formation of new ice. The energy balance of the system is being transformed.

The interaction between Arctic and Atlantic regimes is being played out very clearly in the Barents Sea where the polar front is shifting. If current trends continue, the northern Barents Sea may soon complete the transition from a cold and stratified Arctic to a warm and well-mixed Atlantic-dominated regime that is free of sea ice. The ecological consequences of this *Atlantification* will be profound. Ocean and climate scientists are exploring what the wider impacts may be on weather systems further afield. We explore some of the implications for life in the Arctic in the next chapter.

Chapter 3
Arctic ecosystems

The Arctic biome

Botanists consider the tundra plant communities north of the treeline to be the true Arctic biome. In the treeless landscapes that fringe the Arctic Ocean, the diversity of plants is low, nutrient supply is limited, and soil depth is constrained by permafrost. This biome contains less than 1 per cent of the world's flora and is dominated by hardy communities of low-growing, shallow-rooted shrubs and herbaceous perennials, mosses, and lichen. A distinctive feature of this biome is its high *floristic integrity*—in other words, the flora in a given region will include plants that are found in all parts of the Arctic. Typically, such circumpolar species will account for between 40 and 80 per cent of a given assemblage. This circumpolar integrity has a deep history rooted in the glacial periods of the Quaternary, when Siberia was connected to Alaska via the Bering Land Bridge and millions of large herbivores dispersed seeds, spores, and nutrients across the Arctic landscape.

This chapter aims to capture some of the key characteristics of the Arctic biome, present and past. We will consider how ecosystems function in the northern high latitudes and how they are responding to recent environmental change.

Plants and animals in the Arctic have to cope with extreme conditions. Air temperatures can range from below −40°C in the long dark winters to +20°C in the intense short-lived growing season, but summers are generally very much cooler. The Arctic climate has become increasingly erratic in recent decades with abrupt warmings and regional heatwaves in both winter and summer. Table 2 lists the main stressors for plant growth in the Arctic, many of which relate directly to the harshness of the climate. Arctic plants have a singular trait that distinguishes them from all others—this is the ability to grow, metabolize, and reproduce in air and soil temperatures only slightly above freezing. They exist in what may be called a *thin life zone* between continuous permafrost below and the chilling winds above. Nutrient stocks are minimal because plant material does not break down where soil temperatures are too cold for detritivores and decomposers.

Table 2. Principal environmental stressors for plants in the Arctic

Physical
  Low air and soil temperatures
    Below freezing most of year
    Frosty, or slightly above, during summer
  Very short growing season
  Strong and chilling winds
  Long-lasting snowdrifts
  Permafrost and a shallow active layer
  Thin life zone between permafrost and the wind
  Flooding at spring thaw
  Drought
  Thin and nutrient-deficient soils
  High UV radiation

Biological
  Plant-to-plant competition
  Pollination problems
  Grazing and trampling
  Slow decomposition and nutrient limitation

Modified from Billings, 'Constraints to plant growth, reproduction, and establishment in Arctic environments'.

Arctic vegetation has been grouped into twenty-one provinces based on various characteristics, including relative uniformity of species and number of endemics (Figure 9). The nature of these provinces is intimately related to their Quaternary glacial history, but each contains combinations of lichens, mosses, grasses, sedges, and dwarf woody shrubs. The vast boreal forest ecosystem lies mainly to the south of these provinces, but large areas of coniferous forest underlain by permafrost do lie inside the Arctic Circle in Scandinavia, Alaska, and the far north-west of Canada.

Tiny flowering plants such as purple saxifrage (*Saxifraga oppositifolia*) and the Arctic poppy (*Papaver radicatum*) are found in the sparsely vegetated polar deserts of the High Arctic. Both are found beyond 83° N on Kaffeklubben Island off the northern tip of Greenland—the most northerly land on Earth. The Arctic poppy is a remarkably tough perennial with a circumpolar distribution. It is a heliotrope—its cup-shaped flowers track the movement of the low-angled Arctic sun and offer warmth to pollinating insects. These plants are often found at the bottom of scree slopes and on the margins of gravel-bed streams where rocks return warmth from the sun and offer shelter to plant roots. Precipitation is low across much of the Arctic, but evaporation loss is limited by cool temperatures—snowmelt and summer rainfall are both important for plant growth. The Arctic poppy features on the coat of arms of the territory of Nunavut, northern Canada, alongside several emblematic Arctic species including the caribou (*tuktu*) and narwhal (*qilalugaq tugaalik*). These two species here represent all Arctic animals of land and sea and their importance to Nunavut culture and livelihoods (Figure 10).

Arctic animals

A key feature of the Arctic biome is the high fluctuation in animal populations; this dynamic operates across several timescales depending on species. On a seasonal basis, countless animals come and go by land, sea, and air. Over 200 bird species breed in

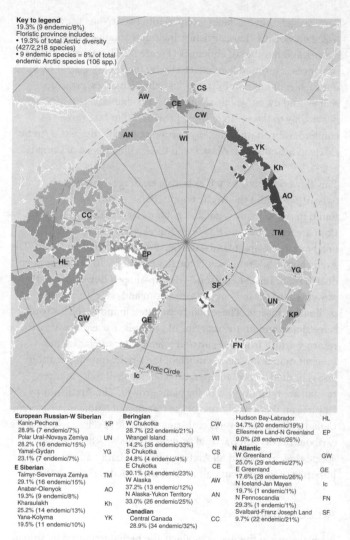

Key to legend
19.3% (9 endemic/8%)
Floristic province includes:
• 19.3% of total Arctic diversity
  (427/2,218 species)
• 9 endemic species = 8% of total
  endemic Arctic species (106 spp.)

European Russian-W Siberian		Beringian		Hudson Bay-Labrador	HL
Kanin-Pechora	KP	W Chukotka	CW	34.7% (20 endemic/19%)	
28.9% (7 endemic/7%)		28.7% (22 endemic/21%)		Ellesmere Land-N Greenland	EP
Polar Ural-Novaya Zemlya	UN	Wrangel Island	WI	9.0% (28 endemic/26%)	
28.2% (16 endemic/15%)		14.2% (35 endemic/33%)		**N Atlantic**	
Yamal-Gydan	YG	S Chukotka	CS	W Greenland	GW
23.1% (7 endemic/7%)		24.8% (4 endemic/4%)		25.0% (29 endemic/27%)	
E Siberian		E Chukotka	CE	E Greenland	GE
Taimyr-Severnaya Zemlya	TM	30.1% (24 endemic/23%)		17.6% (28 endemic/26%)	
29.1% (16 endemic/15%)		W Alaska	AW	N Iceland-Jan Mayen	Ic
Anabar-Olenyok	AO	37.2% (13 endemic/12%)		19.7% (1 endemic/1%)	
19.3% (9 endemic/8%)		N Alaska-Yukon Territory	AN	N Fennoscandia	FN
Kharaulakh	Kh	33.0% (26 endemic/25%)		29.3% (1 endemic/1%)	
25.2% (14 endemic/13%)		**Canadian**		Svalbard-Franz Joseph Land	SF
Yana-Kolyma	YK	Central Canada	CC	9.7% (22 endemic/21%)	
19.5% (11 endemic/10%)		28.9% (34 endemic/32%)			

9. **The Arctic biome north of the treeline showing the floristic diversity (expressed as a percentage of total Arctic diversity, n=2218 species) and proportion of endemics (expressed as a percentage of total Arctic endemics, n=106 species) in twenty-one floristic provinces.**

10. The coat of arms of the territory of Nunavut adopted in 1999, designed by Andrew Qappik from Pangnirtung. See the web link in Further Reading for a full explanation of the images and Inuit text.

the Arctic (Table 3). They account for just 2 per cent of global avian biodiversity. Only the ptarmigans and snow buntings are endemic to the Arctic. The snow bunting (*Plectrophenax nivalis*) is the northernmost breeder among land-based birds with a circumpolar distribution. Breeding in the Arctic has risks because the window is short. A few especially hardy species—including the ptarmigans and auks—are permanent residents and overwinter in the Arctic, although even they will fly south when food is scarce.

Table 3. Selected characteristics of species occurring in the Arctic by taxonomic group

Group	Species occurring in the Arctic	Ratio of worldwide total (%)	Mainly Arctic species	IUCN Endangered, Vulnerable, or Near Threatened	Extinct in modern times
Terrestrial mammals	67	1	18	1	0
Marine mammals	35	27	11	13	1
Terrestrial and freshwater birds	154[a]	2	81[a]	17	0
Marine birds	45[a]	15	24[a]	3	0
Amphibians/reptiles	6	<1	0	0	0
Freshwater and diadromous fishes	127	1	18	0	
Marine fishes	c.250[b]	1	63	4[c]	
Terrestrial and freshwater invertebrates	>4,750				
Marine invertebrates	c.5,000				
Vascular plants	2,218	<1	106[d]	0	0

Bryophytes	c.900	6		
Terrestrial and freshwater algae	>1,700			
Marine algae	>2,300			
Non-lichenized fungi	c.2,030	4		<2%
Lichens	c.1,750	10		c.350
Lichenocolous fungi	373	>20		

[a] Includes only birds that breed in the Arctic.

[b] Excludes the sub-Arctic Bering, Barents, and Norwegian Seas.

[c] Most marine fish species have not been assessed by the International Union for Conservation of Nature (IUCN).

[d] Includes Arctic endemics only.

After Meltofte (ed.), *Arctic Biodiversity Assessment 2013* (see References).

The migration of birds, more than any other animal, links the Arctic, quite literally, to the rest of the world. The remarkable Arctic tern (*Sterna paradisaea*) breeds in the Arctic and flies south to spend a second summer on the Antarctic coast—the longest migration known in the animal kingdom. This little bird sees more daylight than any other animal. In its lifetime it may travel over 2.4 million km. Some species demonstrate remarkable adaptability to the Arctic environment. A study of male great cormorants (*Phalacrocorax carbo*) wintering in west Greenland showed they had dived for fish every day during winter—successfully hunting in the darkness of the polar night.

Arctic animals have evolved a range of adaptations to cope with extreme conditions. The Arctic ground squirrel (*Spermophillus parryii*) hibernates below ground in shallow burrows for eight months (August to April), the longest of any mammal. These creatures draw on fat reserves to sustain an extremely low metabolism while their core body temperature plummets to −3°C. The Arctic ground squirrel prevents its body fluids from freezing by a process known as supercooling—its blood is cleansed of particles that water molecules could form ice crystals around.

Faunal diversity in the Arctic is also rather low with few species of large land mammals and few carnivores surviving the Late Pleistocene extinctions. Table 3 lists selected animal species occurring in the Arctic by taxonomic group and includes eighteen mainly Arctic terrestrial mammals. Note that reindeer and caribou are the same beast (*Rangifer tarandus*): they are called reindeer in Europe and caribou in North America.

Herbivores such as reindeer, woolly mammoth (*Mammuthus primigenius*), and musk oxen (*Ovibos moschatus*) have played a key role in plant community dynamics and nutrient cycling across the Arctic biome for much of the Pleistocene. Research in North America has shown that populations of megafauna fell dramatically well before their final extinctions towards the end of the last ice age. These population declines had a profound and long-lasting impact upon high latitude ecosystems. The removal of herbivores from the landscape allowed wildfires to increase in frequency because plant communities were restructured as woody plants expanded unchecked in a landscape freed from big grazers. Recent work in Alaska suggests that the 'eco-management' work of the woolly mammoths was taken up by other large herbivores including bison, moose, caribou, and musk oxen. These species filled the gap left by the decline of the mammoths, wild horses, and saiga antelopes across large areas of North America.

Today, reindeer and musk oxen are the key large grazers in the Arctic biome. Reindeer forage on a variety of plants including herbs, mosses, and woody species such as willow and birch. In winter, when food is scarce, they scrape away the snow to eat ground lichen. Intensive grazing and trampling by reindeer can cause both plant community biomass and net primary productivity to fall.

One feature of the warming Arctic is an increase in the frequency of winter rain. This can be catastrophic for large herbivores. It can lead to large die-offs because a sudden drop in temperature causes surface icing and prevents access to lichen forage. The Sámi community in Arctic Sweden look after about 8,000 reindeer and have expressed concern about the threat from unusual weather to traditional reindeer feeding areas. Domesticated reindeer are important sources of meat, milk, and hides across the Arctic. On the Yamal Peninsula in the winter of 2013/14, out of a population of 300,000 reindeer, severe ice crusting led to over 60,000 deaths by starvation.

Herd size is important for buffering the impacts of weather and climate variability, but reindeer populations can crash dramatically. When food is scarce, body mass decreases and the smallest individuals are less likely to reproduce. Any offspring are then more prone to starvation and predation. The ecosystem interactions involving large herbivores are complex and harsh winters with limited food supplies can affect herds with a high population density more severely than those that are much lower. In 2010 there were about 3.8 million wild reindeer and 2 million semi-domesticated reindeer in the circumpolar Arctic. The livelihoods of at least twenty indigenous peoples depend on these tundra-dwelling animals.

Connectivity and gene flows

The Arctic fox (*Vulpes lagopus*) is an important carnivore in the Arctic food web. As the only canid with fur-lined paw pads, it is superbly adapted to cold and can survive temperatures below −50°C. Its remarkable endurance in barren polar landscapes has been confirmed by GPS tagging. In spring and summer 2018, a young female fox trekked from Svalbard to Ellesmere Island in northern Canada in just seventy-six days (Figure 11). This intercontinental journey involved a cumulative distance of some 4,415 km between 1 March and 1 July with a leg across the northern edge of the Greenland ice sheet for good measure. She was covering a staggering 155 km per day on the ice sheet—this is the fastest rate of movement recorded for this species. When the Norwegian researchers who conducted the study downloaded the GPS data, they were astonished at both the speed of movement and the distance covered. The northernmost point of the trek was on the sea ice at latitude 84.7° N off the coast of northern Greenland.

Several Arctic carnivores trek long distances when food supplies are low, but this journey was particularly impressive. This fox actually switched ecosystems from the menu of marine resources on offer on Svalbard to Ellesmere Island where foxes mainly prey on brown lemmings. There is extensive gene flow among the Arctic fox populations that are well connected by sea ice. This is an excellent example of circumpolar connectivity between

Movement rate (km/day)
155
0

Ellesmere Island

01 July
10 June
06 June

07-08 April
(stopover)

16 April

10-11 April
(stopover)

Greenland

01 March   26 March

Svalbard

500 km

11. GPS tracking of an Arctic fox journey from Svalbard to Ellesmere Island.

populations and ecosystems; in this case sea ice is critical for the maintenance of genetic diversity and food security, as it can be for polar bears and Inuit too. Sea ice decline, of course, will ultimately result in habitat fragmentation, more isolated populations, and an impoverished gene pool.

Boom and bust: lemmings and snowy owls

Lemmings are the most widespread rodents in the Arctic and a foundational prey species of great ecological importance. They provide food for most Arctic predators including the Arctic fox, snowy owl, and wolverine. There are several species including the Norwegian lemming (*Lemmus lemmus*), the Siberian brown lemming (*Lemmus sibiricus*), and the brown lemming of North America (*Lemmus trimucronatus*). They are important grazers in the Arctic biome, mainly feeding on mosses and grasses. Brown lemmings in Arctic Canada also graze on the flowers, buds, and roots of Arctic willow.

The Norwegian lemming is the only vertebrate endemic to Scandinavia. Lemmings are remarkably hardy creatures, as borne out by their evolutionary history in northern Europe. It was initially believed that the Norwegian lemming colonized Scandinavia from the south and east in the post-glacial period following the decay of the last great ice sheets. Early biogeographical observations, however, suggested that small populations may have survived the rigours of the LGM *within* Scandinavia in local ice-free refugia. This has recently been confirmed by genetic research.

Lemmings can live for up to twenty-four months, but life expectancy is typically less than a year and predation is a key control on their survival. They do not hibernate. During the long Arctic winter lemmings live beneath the snow in insulated burrows lined with mosses and grasses. If conditions are favourable, they can reproduce all year round. Females are fertile at just 3 to 5 weeks of age and produce litters of 6 to 8 young every few weeks. When conditions are favourable, populations can explode dramatically.

Lemming populations show remarkable peaks and crashes over a three- to four-year cycle. The most complete record of Norwegian lemming population changes is from the upland plateau of Hardangervidda close to 60° N where, between 1921 and 2014, there were 26 peaks in the lemming population with a periodicity of 3.6 years. There are two main theories to account for these cycles. The first posits that predators drive the cycles. Populations of the key predators increase as lemmings increase their numbers. Eventually the predators get the upper hand, and the lemming population declines rapidly. The second is the overgrazing theory, whereby the lemming population becomes so large it hoovers up all of the available forage so that large numbers become unsustainable and the population crashes.

Lemmings play a key role in the Arctic ecosystem because their abundance impacts directly on other species. There is a famous photograph of a snowy owl (*Bubo scandiacus*) nest on Bylot Island in the Nunavut Territory lined with over seventy lemming carcasses (Figure 12). Pickings were so rich that summer that the snowy owls killed far more lemmings than they were able to eat. When lemmings are abundant, snowy owl chicks eat well and many more survive. Periodic irruptions of snowy owls across North America have been linked to lemming population peaks when food was plentiful and nesting success was high.

12. A snowy owl nest lined with dead lemmings.

The Arctic melting pot: narwhals and belugas

Rather than isolating populations of the same species, sea ice decline can also bring different species together, raising opportunities for hybridization. Polar bear habitats, for example, overlap increasingly with those of the grizzly bear. In the case of Arctic and near-Arctic marine mammals, dozens of potential hybridizations between discrete populations, species, and genera have been identified. Many of these species are listed as endangered, threatened, or of special concern. This has been described as the Arctic melting pot and a major threat to Arctic biodiversity because rare species in the high latitudes are threatened by extinction.

The beluga whale and the narwhal winter in the Arctic Ocean. Neither has a dorsal fin and both are adapted to swimming beneath sea ice and to great depths. These Arctic endemics are closely related; each has been observed swimming in pods of the other species. Their mating strategies are poorly understood, however, because observing them during the spring sea ice break-up is challenging. Even though these species diverged about 5 million years ago, there has long been speculation over whether or not they could produce viable offspring. For centuries both species have been hunted by indigenous communities—a practice exempt from the International Whaling Commission. The striking helical tusk of the narwhal, which can grow up to 2.5 m in length, is especially prized.

In the late 1980s, an Inuit hunter named Jens Larsen killed three whales in Disko Bay on the western coast of Greenland. Larsen observed that each whale had pectoral flippers like those of belugas and a tail like that of a narwhal. He kept one of the skulls because of its unusual shape. In 1990 the skull ended up in the collections of the Natural History Museum in Copenhagen, where anatomical investigations suggested it might be a hybrid. Almost three decades later, DNA analysis of the skull published in 2019 revealed the specimen to be a male, first-generation hybrid between a female narwhal and a male beluga—the first scientific evidence confirming hybridization between these species. Larsen had killed three *Narluga*.

There is a cautionary tale here. Ecological changes are happening rapidly in all parts of the Arctic on land and in the ocean. As sea ice cover shrinks, species that were once geographically separated can modify their range and new overlaps provide opportunities for species to mate. Infertile hybrids represent a genetic cul-de-sac and a waste of effort. Alternatively, fertile hybrids might thrive in a changing Arctic to outcompete their parent species, whose numbers may already be in decline. Both outcomes are somewhat problematic, but such range shifts and gene flows have always taken place. The large-scale ecological reorganizations currently happening in the Arctic are not unlike those of the glacial–interglacial cycles of the Quaternary ice age. The fossil record tells us that the Arctic tundra biome expanded and contracted throughout the Quaternary ice age and the overlapping of species ranges was quite common.

Arctic greening and browning

Satellite-based observations of Arctic terrestrial ecosystems have reported widespread 'greening' of the landscape over the last four decades (Figure 13). This Arctic-wide increase in biomass has been interpreted as an ecological response to a longer and warmer growing season as well as increased nutrient availability from faster rates of decomposition. The greening of the northern high latitudes is one of the most significant large-scale ecological responses to global climate change. This greening is measured using the normalized difference vegetation index (NDVI) obtained from satellite data. The index is a measure of the density of green cover on an area of land and ranges from −1 to +1. Higher values indicate more dense vegetation cover. Figure 13 shows trends in

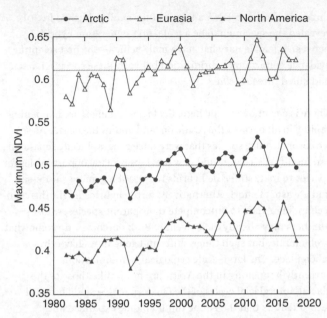

13. **Maximum Normalized Difference Vegetation Index from 1982 to 2018 for the North American Arctic (bottom), Eurasian Arctic (top), and the circumpolar Arctic (middle).**

NDVI—essentially tundra productivity—for different parts of the Arctic over thirty-seven years since 1982. All three curves show significant year on year variability, but the trajectory is steadily upwards in each case.

This greening has also been observed on the ground. In Arctic Canada the increase in plant biomass has largely been driven by the expansion of tall, deciduous shrubs into the tundra biome, especially in wetter habitats. Large-scale 'shrubification' of high latitude ecosystems has the potential to modify the hydrology and the fire regime as well as the foraging opportunities for large herbivores. The 2019 NOAA Arctic Report Card identified the regions with the strongest greening as Alaska's North Slope, the southern tundra of mainland Canada, and the Far East of Siberia. Most tundra soils are strongly nutrient limited and increases in plant growth can only be sustained where faster rates of decomposition enhance the nutrient supply. The deepening of the active layer can provide access to ancient nutrient stores but can also release carbon dioxide, as discussed in Chapter 7.

Arctic tundra vegetation is especially sensitive to changes in summer temperature. Historically, most parts of the High Arctic lowlands have experienced mean temperatures in midsummer a few degrees above freezing. It is important to appreciate that even a small increase in summer air temperature can have a profound influence on biomass and vegetation community structure. This is because during the short growing season under twenty-four-hour sunlight, a small increase in temperature can result in a disproportionate increase in the amount of energy available for plant growth. It is also important to recognize that Arctic greening lumps together a wide range of ecological changes on the ground. These are complex and involve modifications in biomass and plant community structure as well as plant and animal species diversity.

Large herbivores—and how they are managed—exert an important control on tundra plant community structure. In the case of north-western Siberia, which is dominated by semi-managed reindeer herds, large-scale modelling studies have shown that biomass increase due to summer warming is dampened markedly by grazing and trampling by reindeer. So shifts in plant community structure driven by warming can be modified by the action of herbivores such as reindeer and musk oxen. Such interactions are a key feature of Arctic ecosystems and they can modulate plant community responses to changing climate.

The greening trend might be seen as good news, with the potential to capture more carbon in the high latitudes. But the picture is complex and there has been much focus on the ecological consequences for plants and animals in a warmer Arctic. Warming can bring more insects and disease, more extreme weather events and wildfires. All of these are harmful to animals and plants and can lead to the phenomenon of Arctic browning: dead vegetation on a scale that can be mapped by satellite. Some researchers have argued that widespread browning is a signal that Arctic ecosystems are not able to keep pace with the rapidity of recent climate warming. It is the sign of an ecosystem in poor health and a degraded habitat with less food for grazers such as reindeer. Dead and degraded foliage is more liable to wildfires and will be much less effective at capturing carbon and supporting the animals that rely on a healthy tundra ecosystem for their survival. When severe fires burn deep into organic soils and peat, they destroy the insulating layer protecting the permafrost below.

Scientific knowledge is vital to understanding the impacts of climate change on ecosystems, but this must be assimilated with indigenous knowledge to build more resilient Arctic communities. In a study of the impacts of climate warming on ecosystems and traditional livelihoods on Sámi culture in Arctic Finland, researchers from the University of Lapland reported that the most profound negative impacts will be on palsa mire and fell ecosystems, in particular snowbeds, snow patches, and mountain birch forests. A palsa is a peat mound, typically 2–3 m high, with a perennially frozen core of peat or mineral soil. They are widespread in areas of discontinuous permafrost. Ecosystem changes may erode the stories, memories, and traditional knowledge attached to them.

Pleistocene Park

Pleistocene Park is the brainchild of Russian scientist Sergey Zimov, who has ambitious plans for a Siberian Serengeti populated with resurrected woolly mammoths. The woolly part of the plan remains highly unlikely, but it has been successful in generating global publicity for the broader rewilding vision. Vast numbers of big herbivores maintained the ice age tundra steppe ecosystem until overhunting by humans led to their and its demise. Zimov wants to put those beasts back into the landscape. The aim is to upscale a restored ice age grassland ecosystem via a series of connected nature reserves and hundreds of thousands of large animals across the Arctic lowlands of Siberia. The Pleistocene tundra-steppe ecosystem was much more diverse than today's tundra biome, with more species of both large herbivores and predators. The existing megafauna are under threat from the current phase of rapid warming.

Experimental field research inside the Arctic Circle has yielded a radical rewilding plan to combat the effects of climate warming and preserve the Siberian permafrost. This bold idea involves reviving and repopulating the mammoth tundra steppe to replace huge expanses of taiga forest. Grasslands have a higher albedo than forest and will reflect more solar energy thereby helping to keep the soil surface cold. The animals in today's experimental reserve—musk oxen, horses, and bison—keep the grasses short and trample the snow so that the exposed ground, without insulation, freezes hard in winter and keeps greenhouse gases trapped in the permafrost below. Ecological experiments don't normally involve armoured vehicles—Zimov has simulated the tundra-compressing and tree-breaking impact of woolly mammoths by storming across the landscape in a decommissioned Soviet tank.

Range shifts and invasive species

As the Earth warms, it is to be expected that non-native species of plants and animals—on land and in the ocean—will extend their ranges into the Arctic. Ecologists are monitoring range expansions in boreal and temperate species. Mackerel fishing now makes a very significant contribution to the Greenlandic economy even though the first reported catch was only in 2011. The arrival of this fish in Greenland's waters is a remarkable example of how climate change can impact the economy of an entire nation. Fisheries are a major part of Greenland's GDP and play a key part in domestic food consumption. In summer 2019, Greenland proposed to increase its mackerel quota by 18 per cent to 70,000 tonnes per year. This is about half the size of Iceland's quota and has caused tension at the EU. The changing nature of Arctic food webs was further highlighted in 2012 when three endangered Atlantic Bluefin tuna (*Thunnus thynnus*) were caught off Greenland. These tuna prey on mackerel and they are now such frequent visitors to northern latitudes that Iceland and Norway actually set up commercial quotas in 2014.

Apart from some local exceptions, Arctic ecosystems have largely been spared the extreme habitat damage from invasive alien species seen in other biomes. Many animal and plant species and their habitats are unique to the far north, so the Arctic biome possesses very considerable conservation value. Two working groups of the Arctic Council are engaged in strategic planning to combat the threat of alien species. They argue that protecting Arctic ecosystems from alien species may increase their resilience to other threats, most notably climate change. Alien species can be introduced either deliberately or by accident. In the marine environment, for example, as new Arctic shipping lanes see more traffic, alien species can be introduced via the discharge of ballast water. Earthworms are reaching the Arctic via imported soils for gardening and in the treads of travellers' boots. They are remarkable ecosystem engineers. Experiments have shown that adding earthworms to Arctic soils can markedly enhance nutrient cycling, leading to increases in tundra heath and meadow biomass equivalent to a 3°C rise in air temperature. Under their own steam, as soils warm, earthworms are moving northwards at a speed of 5–10 m a year.

Perhaps the most notorious example of a deliberate release is the case of the red king crab (*Paralithodes camtschaticus*) in the Barents Sea (Figure 14). A native of the far northern waters of the Pacific, it was introduced to the Barents Sea in the 1960s by Soviet

14. The invasive giant red king crab is now a key part of the marine ecosystem off northern Norway.

scientists seeking to create a new commercial fishing resource. A highly valued delicacy, it can weigh over 12 kilos with a leg span of 1.8 m—it is the largest of the king crabs. This Arctic invader has adapted spectacularly to its new habitat and millions of individuals have spread westwards to the fjords of Norway as far as Troms. The red king crab has extended its range westwards by about 15 km each year. It has few natural enemies and eats everything from cod larvae to other crabs and commercial scallop beds. By 2006 there were upwards of 20 million individuals in the Barents Sea. Ecologists have predicted it will reach Svalbard by 2030.

Arctic food webs

Arctic food webs are typically dominated by a few key species and this makes them vulnerable to environmental change. When the climate warms at the top of the world cold-adapted species have nowhere to go because cold refuges get too small to support viable populations—this leads to ecological collapse and extinction. Apex predators like polar bears and narwhals are having to cover greater distances and use much more energy to survive. It has been predicted that starvation and reproductive failure will drive the polar bear to the brink of extinction by the end of this century.

The decline in sea ice extent has far-reaching consequences. A National Oceanic and Atmospheric Administration (NOAA) study between 2007 and 2018 found that three species of Arctic seals in the Bering Sea and the Aleutian Islands were losing body mass in almost all age and sex classes. The decline in the ribbon and spotted seals was attributed to their reliance on the sea ice during pregnancy and nursing cycles when the energy needs of mothers are highest. The fall in sea ice extent has forced mothers to search for food in less favourable areas.

Because terrestrial ecosystems in the Arctic are adapted to low-temperature regimes, they are especially vulnerable to rapid climate warming. One of the key drivers of change in Arctic ecosystems—on land and in the ocean—will be the later arrival of winter and the earlier arrival of spring. These variables may be more important than an overall increase in mean air temperatures because they influence the timing of animal reproduction and migration and therefore breeding success.

The Arctic Ocean and its shallow seas host over 5,000 animal species and contain commercially important fisheries. The Arctic char (*Salvelinus alpinus*) is the world's most northerly freshwater fish—a cold-water-adapted salmonid with a circumpolar distribution that is notable for a high number of morphological variants. It is found in a range of habitats including lakes and rivers and some populations are anadromous—they spawn in rivers and then migrate to the sea. A recent study of Arctic char populations along the west coast of Greenland identified potentially important adaptations to geographical variations in sea surface temperature and the best time of year for migrating to sea. Gene flow among these populations may allow a degree of adaptation to a warming ocean.

Arctic ecosystems are full of surprises. The Greenland shark, for example, can live for more than 400 years—it is the oldest living vertebrate, yet we know very little about its ecology and how it might cope in a warmer world. Spitsbergen's bowhead whales (*Balaena mysticetus*) were almost wiped out by the world's earliest commercial whaling operations that began in the early years of the 17th century. In the 1990s they were thought to be on the brink of extinction. The bowhead is the only baleen whale that spends all year in the Arctic. They feed on Arctic crustaceans and spend the winter in deep frigid waters almost completely covered in sea ice. It has been suggested that commercial whaling wiped out the bowhead populations that spent the winter further south, whilst those that wintered in the north amongst the sea ice evaded the whaling ships. The descendants of these hardy, cold-loving populations are now growing in number and their songs have recently been detected by Norwegian researchers throughout the deep waters between Greenland and Svalbard. This is a heart-warming story of survival but there is much concern about the prospects for these cold-loving whales in a warming Arctic.

Whether ice, land, or water, High Arctic ecosystems can form rich wildlife habitats. One of the key challenges for better understanding these ecosystems and how they may respond to climate change is the fragmentary nature of many datasets and the lack of long-term records of species behaviour. Whether studying ice algae or polar bears, there are big gaps in our knowledge and this environment poses formidable logistical challenges. Animals in the Arctic are tuned to the seasonal rhythms of temperature that tell species when to mate or when to migrate. As the climate warms, the thermal cues are shifting and with unknown consequences. Many changes are happening too rapidly for us to fully understand the cascade of ecological consequences. But there is a great deal of innovative research in progress and much international collaboration. Big data studies utilizing GPS tagging and satellite observations are gathering information on how Arctic species are responding to a warming climate. The Arctic Animal Movement Archive (AAMA), for example, is a database of over 200 tracking studies of terrestrial and marine animals from 1991 onwards. It has shown, for example, how golden eagles have been starting their spring migration to the Arctic half a day earlier each year over the past twenty-five years. Female reindeer are migrating earlier to give birth—this is risky because it can increase the chances of vulnerable offspring encountering freak blizzards. Autonomous submarines are gathering data on the processes associated with seawater temperature changes so that we can better understand the impacts on nutrient supplies and the functioning of marine food webs. Research in Greenland fjords has indicated that Arctic food webs can be quite resilient. It remains to be seen how long this will be maintained given the pace of Arctic warming on land and in the ocean.

Chapter 4
Peoples of the Arctic

Four million people live north of the Arctic Circle and approximately 2 million of them are found in Russia. An important distinction should be made between indigenous and settler residents since the latter outnumber the former 3 to 1 (Figure 15). Just over 1 million indigenous people live in the northern territories of the eight Arctic states. This includes: 8,100 Aleuts in Alaska and the Russian north; 32,400 Athabascans in Alaska and northern Canada; around 145,900 Inuit in Alaska, northern Canada, and Greenland; approximately 78,000 Sámi in northern Norway, Sweden, Finland, and Russia; and 866,400 people in 'northern Russia' belonging to a variety of indigenous groups. In the case of Russia, the indigenous population north of the Arctic Circle is closer to 260,000 (Table 4).

The balance between indigenous and settler varies greatly from place to place. In Greenland, well over 90 per cent of the island's 56,000 population identify as indigenous. In the vast Canadian north, there are only 130,000 residents scattered across the three territories of Northwest Territories, Yukon, and Nunavut and indigenous peoples make up about 50 per cent of that total. These figures and those listed in Table 4 are invariably controversial and should be treated with caution. Census data for Canada, Russia, and the United States do record indigeneity, whereas in Greenland

Table 4. Some characteristics of Arctic states

Country	Canada	Finland	Greenland (Denmark)	Iceland	Norway	Sweden	Russia	Alaska (USA)
'Arctic' population (estimated numbers)	113,000	179,000	56,000	350,000	393,000	1.2 million	2 million	737,000
Indigenous population (estimated numbers)	58,000	8,000	56,000	Non-applicable	50,000	20,000	>250,000	86,000
Approximate size of Arctic or 'northern' territory (% of national territory)	40	30	66	100	35 (excluding Svalbard)	15	30	17
NATO membership	Yes	No	Yes	Yes	Yes	No	No	Yes

Arctic ocean coastline	Yes	No	Yes	No	Yes	No	Yes	No
Permafrost	Yes	Yes	Yes	No	Yes	Yes	Yes	Yes
Main industries in Arctic territories	Mining, fishing, tourism, energy	Forestry, tourism, mining	Fishing, tourism, and mining	Fishing, tourism, renewable energy, smelting	Oil, gas, fishing, tourism, timber	Forestry, mining, tourism	Oil, gas, mining, timber,	Seafood, oil, gas, timber, mining, tourism

Note: Population estimates in this table are based on available data for the last few years from a range of sources. Compare with data for the first decade of this century in Figure 15.

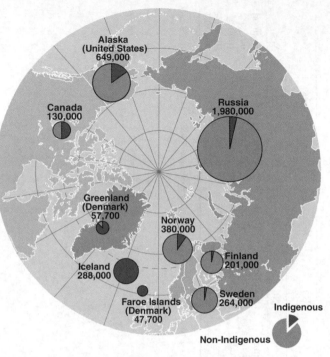

15. Indigenous and non-indigenous populations in the Arctic. These data are estimates for the first decade of the 21st century.

and Scandinavia other measures might be used, including place of birth and voter records.

Historically, Arctic states have also either failed to accurately capture population numbers and/or actively sought to downplay the number and significance of indigenous peoples, with attempts made to de-indigenize communities, through the use of residential schools and active prohibition of indigenous languages. Given the political and cultural sensitivities attached to population counting and census reporting, many researchers working with indigenous communities suggest that one should focus on individual and community self-identity (which could be multiple) rather than ancestry or place of birth.

Traditional indigenous family structures vary across the Arctic. There are nine Inuit groups in Canada alone. In Inuit cultures, for example, families lived and hunted together in communities that might number thirty or more people depending on the season. Inuit communities tended to be larger in the winter months than summer where communities moved around in search of food. Marriage could be either arranged or by choice. Gender divisions of labour were the norm. Men tended to be hunters and fishers and women tended to look after children, manage households, and make clothing out of plants and animal skins. Community leaders were traditionally men and often selected on the basis of their hunting skills.

Foods such as seal and whale are commonly shared widely amongst local communities, and such sharing rituals are considered integral to social cohesion. What surveys and census data have tended to reveal is that multi-generational family living remains widespread with implications for overcrowding alongside poor mental health and physical well-being. While there remain structural challenges facing many Arctic communities, it is important to recognize that Inuit and other indigenous peoples have been actively involved in generating their own educational resources and designing innovative health and educational strategies that are sensitive to the learning needs of their communities. The annual Arctic Inspiration Prizes in Canada provide examples of such innovation, including a 2021 prize winner involving the mentoring and teaching of Inuit children in indigenous music and dance.

Settler communities in the Arctic are also highly varied: from city residents north and just south of the Arctic Circle, such as those in Anchorage and Murmansk in Alaska and Russia respectively, to contract workers who might 'fly in and fly out' from resource development sites across the Arctic. In the Russian north, one might find migrant workers from the Central Asian republics and Turkey working on liquefied gas projects. Indigenous and settler communities live and work together in villages and towns across the Arctic. Given the strategic importance of the Arctic to all the Arctic nations, military personnel and their families are to be found at numerous air force bases, naval ports and harbours, and army facilities. If you fly up to Alaska, for example, it is commonplace for Alaska Airlines to announce to passengers that they should remain in their seats so that any service personnel can depart first. The US armed forces and the defence community are hugely important to the Alaskan economy. At any one time, there might be 25,000 to 30,000 people in Alaska connected to the armed forces, National Guard, reserves, and civilian defence community, who in turn help to support local businesses and indigenous/native corporations. There is no one Arctic community per se, and the indigenous peoples of the Arctic live and work in 'southern' cities such as Copenhagen, Toronto, Oslo, and Moscow.

Arctic migrations

The first people to arrive in the Arctic probably did so some 30,000 to 40,000 years ago. Archaeological evidence suggests that there were Upper Palaeolithic hunting communities in and around the Yana River in north-eastern Siberia. These people hunted woolly mammoth and other large animals such as woolly rhinoceros and bison. The Yana people settled in an area of the Euro-Asian landmass that was largely ice free. What we do know about the earliest migration and settlement is inevitably fragmentary and well-dated sites are few in number. Indigenous storytelling enriches this archaeological accounting of Arctic prehistory with Inuit believing that a mysterious ancient people named Tunit called the Arctic home before their arrival. From north-east Siberia humans migrated across the Bering Land Bridge to North America and did so between about 16,000 and 12,000 years ago. Many scholars believe that the coastal route was the most important. Some migration might even have occurred before that time, but more robust dating of the archaeology is needed to make this case. In 1937 the Swedish botanist Eric Hulten coined the term Beringia to refer to a cultural-ecological zone, encompassing eastern Siberia, Alaska, and north-west Canada. The migratory movements of the earliest settlers were inevitably shaped by where it was possible to hunt and survive in a world still in the grip of an immense glaciation. Not everyone crossed the land bridge, however, with some groups remaining as Palaeo-Siberian communities and moving inland again to mix with other groups living in the modern-day Russian Far East. Climate change may have been one driver of human migration, but it is likely that there were other factors that drove changes to social structures and the geography of communities. Humans followed animals as they migrated, and they exploited marine resources along the coast. Although much evidence must now be below sea level, there is good evidence to suggest that the route along the coast and islands of north-east Canada was important.

At the end of the Pleistocene ice age, sea level rose rapidly and low-lying regions such as the Bering Land Bridge were submerged. It is believed that the land bridge disappeared around 11,000 years before present (BP). What we now call the Bering Strait separates the Asian and North American continents. Over time, the earliest settlers, the Palaeo-Arctic peoples, began to migrate southwards and eastwards as ice retreat opened up river valleys and plains. But what cannot be stressed enough is that there is plenty of disagreement about the timings of these trans-continental movements, the routes taken, and the source and numbers of people involved. Did the earliest migrants travel by land corridor or did some travel along the coast by boat? The story of the trans-Bering pioneers will continue to be enriched and informed by archaeological research. What is considered most likely is that there were additional migrations shaping the settlement of the North American Arctic from around 9,000 years BP onwards: a Palaeo-Inuit people hailing from the far east of

Siberia migrated across the Bering Land Bridge; a Thule people who were whale-hunters in northern Canada and Greenland established a 'Dorset Culture' with ice-based seal hunting and igloos; all of which contributed to a further diversification of the Arctic around 4,500 years ago, with small communities engaged in whaling, sealing, and inland hunting.

The fate of communities in the Arctic is explained in part by environmental change, the availability of marine resources, and technological change. Further migrations involving Eskimo-Aleut and Na-Dene-speaking peoples saw more communities heading south towards what are now the US–Canadian borderlands. Modern research, using the DNA of Inuit and other native peoples, has focused on identifying waves of migration and discredited earlier theories which tend to focus attention on one large migration event and identification of the First Americans as so-called Clovis peoples. What we are learning is that Arctic peoples in North America have genetic heritage pointing to multiple migrations from Siberia to the Americas and that people also moved in the opposite direction. Around 2,800 years ago, the archaeological evidence for Palaeo-Inuit culture trails off, suggestive of population replacement by the ancestors of contemporary Arctic peoples such as Inuit and Yupik.

The next big wave of Arctic settlement occurred far later, involving Norse exploration and settlement of southern Greenland. In the 10th century, Norse settlements were established with a network of farms and churches hosting a community of around 6,000 people. What the Icelandic explorer Erik the Red termed 'Greenland' proved sufficiently attractive to a swathe of fellow colonialists (Figure 16). This community was believed to have existed for around 500 years until a cooling climate was held responsible for making southern Greenland less attractive for farming and fishing.

Until quite recently, ideas around the peopling of the Arctic were largely derived from evidence about environment and climate.

16. Norwegian Viking Erik the Red in Greenland encountering the landscapes and creatures of the Arctic. The image was painted by Danish artist Jens Erik Carl Rasmussen (1841–93), who visited Greenland in 1870–1. On a return visit in 1893 he fell overboard and drowned near the Shetland Islands. While Norway became independent from Denmark in 1814, Rasmussen's painting is a reminder of a shared Nordic tradition of exploration and discovery.

This has been pivotal to making sense of the changing fortunes of indigenous and settler communities. The fate of the Norse in Greenland remains the most notable. Why did the Norse die out in Greenland in the 15th century after around 500 years of living and trading in the south of the island? Newer research is suggesting that the decline of the Norse settlements in Greenland might have been due to a gamut of reasons, including one theory that Norse traders were being 'frozen out' of markets for walrus ivory, fur, and pelts. In other words, a colder climate would certainly have made life harder in both ice-covered fishing grounds and snow-covered sheep farms, but these communities were not simply declining because of environmental pressures.

Ivory markets in Europe were saturated and the once lucrative trade in walrus ivory was becoming ever more precarious. By the 15th century, European consumers no longer considered walrus ivory to be fashionable and this loss of revenue was disastrous to distant Norse communities.

As Europeans started to explore, exploit, and settle the Arctic from the 16th and 17th centuries onwards, new waves of migration began to impact upon Arctic regions and peoples. While walrus ivory was in decline, other industries such as whaling and fishing continued, alongside the relentless search for new trade routes around the Americas and Asia. All of this prompted settlers and speculators to establish new stations and camps across the Arctic, including on remote archipelagos such as Svalbard. By the 17th century, mining was transforming northern Scandinavia, which, like the northern parts of North America, had witnessed settlement and exploitation dating back millennia. In more recent times, the exploitation of copper, gold, iron ore, and zinc sparked further settlement and development in Sweden, Greenland, Svalbard, and further afield in North America, where immigrant labour from China played an important role in enabling 'gold rushes', railway building, and fish processing. Later on oil- and gas-related development across the Arctic fuelled additional population growth, as areas such as the North Slope in Alaska attracted inward investment and infrastructural development. The next chapter provides further reflection on the scale and pace of resource development and exploitation.

Indigenous Arctic peoples

Indigenous peoples have called the Arctic home for millennia. The land and waters of the Arctic are integral to those communities (Figure 17). Among those who have called the Arctic home, there is considerable diversity. In Russia and northern Scandinavia, the Nenets and Sámi are traditionally reindeer herders, their nomadic lifestyles shaped by the vicissitudes

17. An Inuit family sewing seal skin onto the wooden frame of a kayak in Greenland. The kayak is supported on two barrels. A seal skin tent is behind, rocky terrain and water with ice floes in the background.

of the short summer and long winter seasons. In Alaska, northern Canada, and Greenland, Inuit live on the coastline and have organized their lives around sustainably hunting marine mammals such as seals, whales, and walrus, and other animals such as caribou and musk oxen. The coastal waters of the Arctic are their 'marine garden' in the absence of cultivatable land.

Four things unify the indigenous peoples of the Arctic. First, this connection to land and water is profound. It not only provides essential resources for survival but is also integral to community self-identity and cosmology. The belief-systems and oral storytelling cultures of indigenous societies are rooted in experience and respect for the land, sea, sky, and ice of the Arctic. Indigenous cosmologies do vary across the Arctic, with Inuit believing that the Earth was created by a Great Raven who dropped a huge rock, which created the islands and continents of the world. In the Scandinavian Arctic, creation myths centre around the role of giants and goddesses. While reindeer, seals, and whales inform those cosmologies, they also provide food and resources to produce clothing and equipment such as gloves and spearheads. The material cultures of Arctic peoples are rich and varied. In the Siberian Arctic, archaeologists have discovered mammoth ivory needles dating to around 30,000 years ago. Those needles would have been used to make clothing, tents, and bags, including waterproof suits of seal hide which insulated hunters from the frigid waters of the Arctic. Wood bark, grasses, and fish gut have all been used to make things in the past and present. Few items are wasted by Arctic communities whose lives depended on their relationship to land and sea.

Second, indigenous peoples live and work with ice and a prevailing cold climate in ways that are relentlessly attentive to seasonality. Cold temperatures and strong winds help to dry and preserve food all year around. A short summer season provides opportunities to pick berries, to hunt and travel further afield, and collect grasses and bark for clothing and basket manufacture. Winters in the Arctic are cold and lengthy, but ice and snow allows for mobility between and beyond villages, via sledges, skis, and snowmobiles. Ice can be used for travel, as a building material, and to preserve food. Third, indigenous cultures were fundamentally oral and later developed written language, which had important implications for Arctic peoples when they encountered external parties. Finally, all Arctic indigenous communities have had to not only battle with changing climates and environments, but also endure contact with outsiders.

All of this has brought considerable costs and disruption to nomadic and subsistence lifestyles. As northern territories were colonized by imperial adventurers, trading companies, and nation-states, indigenous communities were exposed to disease, co-opted into resource exploitation, marginalized, and even removed from their traditional homelands. More recent attention has focused on the impact of climate change—how to continue to live with the land, ice, and sea when all around is changing in the form of more heat, less sea ice, thawing ground, and now, most recently, so-called zombie fires. Less sea ice, for example, means that Arctic coastal communities are more exposed to violent winter storms. Landfast sea ice acted as a reliable buffer between ocean and land. Thawing permafrost causes ground instability, which makes buildings unstable, even untenable. Frozen ground acts as a natural freezer for harvested food such as whale and seal. Thawing ground imperils that. It is also important to acknowledge that there is rapidly growing external interest, as later chapters note, in the mineral resources and strategic importance of the Arctic.

The intergovernmental body, the Arctic Council, recognizes six permanent participants (indigenous Arctic organizations), and these reveal the linguistic and cultural diversity of Arctic peoples. The Aleut International Association (established in 1988) represents the interests of Aleut people who live along the shoreline and islands in and around the Bering Sea, in Alaska and the eastern edges of the Russian Federation. Numbering around 20,000 people, Aleut speak Russian and English as well as the Aleut language (with two distinct dialects, Eastern and Atkan), which is endangered. Aleut peoples have been active in highlighting the environmental and political challenges facing their communities. The Bering Sea Elders Group is an association representing thirty-nine tribes in the Yukon-Kuskokwim and Bering Strait regions of Alaska who advise and support their work. The elders have warned about the disruptive effects of commercial shipping and fishing to local marine ecologies and community resilience. In 2019, the US agency NOAA invited the indigenous communities around the Bering Sea to inform the Arctic Report Card with a first-hand account of how commercial and environmental change was affecting local communities.

Further north and east, the Arctic Athabascan Council (created in 2000) represents 45,000 people who live in Alaska, Yukon, and Northwest Territories in Canada. Whereas the Aleut traditionally live on coastlines and islands, Athabascan peoples made their homes on tundra and taiga. The Dene-Athabascan language family (including Gwich'in) is not confined to the north, however. Athabascan speakers are to be found all over the Pacific coast of Canada and the Lower 48 of the United States. Notably in 2013, the Arctic Athabascan Council (AAC) filed a petition to the Inter-American Commission on Human Rights, noting the damage being inflicted upon their communities by emissions of black carbon from Canada. Although not unique (there was an earlier petition by Inuit in 2005 regarding the United States), it provides a good example of how indigenous communities are demanding respect for their human rights and insisting upon consultation and engagement in the face of continuing environmental change, resource development, and commercial activity such as shipping and fishing. In early 2021 the AAC was still awaiting a judgment from the Inter-American Commission on Human Rights.

Alongside Athabascan and Aleut peoples, there are around 9,000 Gwich'in who live in small communities in Alaska, Northwest Territories, and Yukon. Represented by the Gwich'in Council International (established in 1999), the community has traditionally been closely associated with the Porcupine Caribou (wild reindeer) herd. For a thousand years, Gwich'in have harvested the caribou, and this has provided an essential food resource as well as materials for clothing and other objects. Unlike in Europe and Russia, where reindeer were domesticated and kept in herds, caribou have always been hunted. As highly migratory animals, caribou criss-cross the US–Canadian border: their breeding grounds include the Arctic National Wildlife Refuge, which has been coveted by former administrations as a potential site for oil and gas exploration.

Formed in 1977, the Inuit Circumpolar Council (ICC) is the most significant permanent participant of the Arctic Council. Representing some 160,000 Inuit who live in Alaska, Canada, Greenland, and Russia (Yupik peoples), the ICC has been at the forefront of campaigning for indigenous Arctic peoples. Apart from English, Danish, French (some Inuit live in northern Quebec), and Russian, the Inuit language is called Inuktitut in Canada, and Kalaallisut, which is the official language of Greenland. Inuit are traditionally a hunting culture and have harvested marine mammals such as seal and whale as well as other food resources on land such as caribou. Inuit were adversely affected by export bans on seal products imposed by the EU and the United States, and the ICC highlighted how climate change, geopolitics, environmental conservation measures, and trade restrictions are making Inuit communities insecure.

In Russia, the Russian Association of Indigenous Peoples of the North (RAIPON, created in 1990) represents the forty groups who live across the vast northern territories of the Federation. With a combined population of over 250,000, RAIPON's members occupy some 60 per cent of the country's territory. It has campaigned on issues such as human rights and protested against forestry and energy development projects that frequently ride roughshod over indigenous and community wishes and rights. In the recent past, Russian authorities have shown a willingness to shut down RAIPON, accusing the organization of being infiltrated by 'foreign agents'. The languages spoken in the Russian north reflect the wider diversity and mobility of indigenous communities living in Russia's circumpolar territories, connecting local groups such as Nenets, Even, and Sakha peoples to the Manchu-Tungus, Turkic, and Palaeo-Siberian language families.

Finally, there is the Sámi Council (established in 1956), the oldest indigenous Arctic peoples' organization, which promotes the interests and rights of Sámi living in Norway, Sweden, Finland,

and the Russian Kola Peninsula. It is estimated there are around 100,000 Sámi living across the four countries. In the Nordic Arctic, Sámi have far greater rights and representation including parliaments. Sámi are traditional reindeer herders and like RAIPON have had to struggle to protect local languages, access to land and water, and cultural recognition. The Sámi language is related to the wider Uralic language family, which includes Finnish and Hungarian.

Future of Arctic peoples

The human settlement of the Arctic has a deep history. Over thousands of years, people have moved in and out of the Arctic region. Two key factors have made human migration in the far north distinctive: adaptation to the cold climate and physical distance between population centres and ever smaller settlements. Indigenous communities in the Arctic have had to be versatile and highly adaptive to their environments, which in practice meant understanding well the seasonality of weather and food availability, and careful use of resources, including those essential for clothing and shelter. Over millennia, a combination of population pressures, climate change, resource availability, and economic factors have affected Arctic communities.

Looking ahead there is no doubt that we will continue to see change, but we should take care not to ascribe all of this to environmental change. The number and distribution of people in the Arctic is down to an amalgam of factors. For instance, Arctic states have long encouraged nomadic peoples to live in settlements, which makes it easier and more cost-effective to provide essential services and education. Arctic peoples have not only been forced to settle into communities, but also relocated to other parts of the Arctic. In both cases, this was motivated by a desire of Arctic states such as Russia and Canada to exert further control over their national territories. Arctic peoples have also migrated away from the Arctic in search of employment and educational opportunities. While climate change is providing yet another driver of change, geopolitical and geo-economic factors (such as the global price of oil and gas) also play their part in shaping the peopling of the Arctic.

The Arctic's human population, despite climate change, is likely to remain at around 4 million in the coming decades. Arctic demographers predict that more and more Arctic peoples will be based in towns and cities, but in the Russian and North American Arctic (including Greenland) there will still be dispersed and small-scale settlements. Infrastructure provision will vary even if more of those communities begin to enjoy the benefits of better internet access. The Russian north is thought to be likely to lose population, in part because of declining birth rates and outward migration. Another noticeable feature of the Russian Arctic is the shortage of men, many of whom die prematurely due to disease and illness. Some parts of the Arctic have historically had more

18. Nuuk, 21 June 2009: Denmark's Queen Margrethe hands over the legal document pertaining to Self-Rule to the chairman of the Greenlandic Parliament, Josef Motzfeldt, during an official ceremony marking the new era of Greenlandic Self-Rule. 21 June is Greenland's National Day.

men than women, and this reflects the labour demand of mining industries and the military, for example. Alaska would be the prime example of how the peopling of the Arctic has been shaped by resource booms and conflict including the long Cold War. As Greenland moves closer to independence (Figure 18), it is quite possible we will see less in-migration from Denmark as Danes and the Danish language lose their influence. In the Canadian Arctic, the territory of Nunavut is one area of the Arctic experiencing population growth due to higher birth rates, especially amongst indigenous families. There is, in short, considerable variation in demographic trends within the Arctic.

Chapter 5
Exploration and exploitation

The history of Arctic exploration and exploitation owes a great deal to early European encounters with the 'New World', from the earliest Viking settlement of Greenland to a succession of European explorers and expeditions designed to search for what they termed the Northwest Passage—the fabled sailing route between Europe and Asia. Viking settlements were established in Iceland by the 9th century and trading and warring with Inuit in southern Greenland was chronicled in the 10th century. Viking exploration continued apace after the 10th century and extended north-east towards the Kola Peninsula and the Barents Sea. Vikings traded with indigenous Arctic peoples in northern Scandinavia. Whale ivory, seal pelts, fish, and animal furs were being exchanged and traded all over the Arctic and northern Europe. By the 13th century, a diverse community of traders, Christian missionaries, and pioneering settlers was operating across the North American Arctic, including Greenland, as well as the Russian and Scandinavian Arctic. Viking settlers developed their farms and established trading networks with other parts of Europe. Coinciding with a period of relatively warmer climate, Viking trade was enabled by more favourable sea ice conditions.

By the time European explorers in the early modern and modern period (c.15th to 19th centuries) began to explore, trade, survey, and exploit marine resources, indigenous peoples and settlers

19. An Iñupiat hunter on the Point Hope headland, Alaska in 1900. He sits on a whale that has been hauled on to the sea ice for butchering.

such as the Vikings had a long-established reservoir of knowledge of these northern territories including their coastlines, sea ice, and open water (Figure 19). Notwithstanding the claims made by European sailors and expeditioners, the Arctic had been lived in for thousands of years. It was not a blank space as European maps postulated.

What the Columbian era brought to the Arctic was more exploration, resource exploitation, settlement, and ultimately the displacement of indigenous peoples by modern settler colonial states. It also brought disease and societal disruption, experienced by native peoples around the world. The High North was seen by many interlopers as an inexhaustible resource frontier—a land of plenty—filled with timber, minerals, furs, fish, ivory, and precious metals. The Canadian oil industry started in Norman Wells in the

Northwest Territories in 1920. While Imperial Oil Ltd sank the first well, natural oil seepages had been reported by European travellers as early as the 18th century. English and Dutch whalers were operating around Svalbard in the 17th century and copper mining in Falun in northern Sweden goes back to medieval times. There is a rich industrial archaeology of abandoned mines and whaling stations across the Arctic. The most dramatic examples of resource development are to be found in the Soviet Arctic where heavily industrialized, specialist 'resource cities' were built to exploit the mineral wealth. The 'Red Arctic' was integral to Soviet economic development and geopolitical power. An early project was the Stalinist era White Sea–Baltic Canal which opened in 1933 after twenty months of forced labour. Between the late 1940s and early 1950s, thousands of political prisoners constructed an Arctic railway connecting nickel-producing areas in Siberia to factories in the Urals. It became known as the 'Arctic death road' and was abandoned after Stalin's death in 1953. Stalin's railway ambitions were resurrected in 2019: The Transpolar Mainline project aims to reconnect the railway from the Ural Mountains to the resource city of Norilsk and then to resource cities and Liquified Natural Gas (LNG) terminals on the Arctic Russian coast.

Exploration and exploitation continue to go hand in hand in the Arctic. Resource development and exploitation supports and enables indigenous Arctic economies and societies. It provides employment and in recent decades native communities in the north have acquired greater rights over land and resource development. From Alaska and Yukon to Greenland, local peoples are now far more involved in decision-making about mining and other projects. While environmental organizations might be eager to terminate resource and energy projects, this sentiment is not shared uncritically in the north. Mining projects not only bring the promise of extra infrastructural investment in roads, digital networks, airports, and ports, they also place further pressure on 'southern' governments not to neglect their northern constituencies. This is noticeable when foreign investors are involved. When Chinese investment is discussed in Canada and Greenland, it unleashes concerns that a coalition of indigenous peoples and Chinese investors will combine to enable large-scale investment and extraction of minerals. National governments in Ottawa, Copenhagen, and Washington DC might then find themselves sidelined because of their tardiness to invest in the north or caught up in struggles to reconcile resource development with climate change commitments and low carbon transition planning.

The Arctic is ground zero for all the contradictions that humanity will face in this century. Sámi peoples in the Scandinavian north are locked into disputes with energy developers and national governments about wind farms, gold mining, and supporting infrastructure, which interferes with reindeer herding. Some of the eight Arctic states continue to invest heavily in oil and gas development in the offshore Arctic at the same time as the renewable energy sector in the form of hydro and wind power is expanding. All the Arctic states worry about the cumulative effects of permafrost thaw, wildfires, and sea ice loss, but they still think of the Arctic as a resource zone. Russia and Canada remain resource-based economies, and their northern territories are vast. China is investing in and working with both countries, including indigenous peoples. In 2020, the Canadian government rejected the sale of an ailing mining company called TMAC Resources, which operates in Hope Bay, Nunavut. The would-be buyer was the Chinese state-owned Shandong Gold Company. The strategic relationship between Russia and China will prevail given the effects of sanctions imposed on Russia in the post-Crimea annexation era. Indigenous leaders across the Arctic have been clear that they are looking for investment and employment opportunities, and China represents an opportunity not a threat.

After the Vikings

From the 15th century onwards, there were repeated hopes that a shorter shipping route would be found between Europe and Asia,

and thus bypass the Ottoman Empire, which dominated
land-based trade routes between Europe and Asia. Map-makers
such as the Flemish cartographer Gerardus Mercator (1512–94)
played their part in filling in the 'blank spots' on European maps
of the world (Figure 20). Dreams of discovering a Northwest
Passage were often dashed by the harsh geographical and
climatological realities. The passage itself spans some 900 nautical
miles from Baffin Island on the North Atlantic coastline to the
Beaufort Sea, lying off modern-day Alaska. Any ship wishing to sail
between those two points had to navigate freezing weather, cold
and ice-filled waters, huge icebergs, and multiple islands. Despite
speculation that there might be 'open polar seas', sea ice remained
an enduring barrier to this route for hundreds of years.

20. Gerardus Mercator's second draft of the *Septentrionalium
Terrarum* (1606).

One of the earliest explorers to try to find the Northwest Passage was a Venetian navigator called John Cabot (about 1450–99). Setting off from Bristol in England in the spring of 1497, he reached the Canadian North Atlantic coastline in May. A year later Henry VII gave his support for a second expedition. Unfortunately, the five expeditionary ships were consumed by a violent storm in the North Atlantic and never made it back to Bristol. In the 17th century, Henry Hudson (1565–1611), who was employed by the Dutch East India Company, led an expedition towards North America in search of the Northwest Passage. The end result was that Hudson visited Long Island, gave his name to the Hudson River, and enabled the Dutch colonization and settlement of New Amsterdam (later New York). He journeyed again towards the Americas in 1610 and entered what was to be later called Hudson's Bay (now Hudson Bay). Unfortunately for his crew, the ships were trapped in sea ice and a mutiny occurred. Hudson disappeared in murky circumstances.

The most notorious expedition to search for the Northwest Passage was that led by the British explorer Sir John Franklin. Accompanied by around 130 men on board the ultra-modern ships HMS *Erebus* and *Terror*, the expedition departed in 1845. After encountering horrendous weather and sea ice, all contact ceased. The ships had become trapped in the sea ice (Figure 21). As their supplies dwindled, the crews attempted to escape on foot. Reports from indigenous people suggested they might have resorted to cannibalism as the men starved and succumbed to the intense cold. For years afterwards, Sir John's widow Lady Jane Franklin urged governments and sponsors to launch their own expeditions to find Franklin and his men. In the 1850s, the Irish explorer Robert McClure was the first to uncover evidence of how the men must have left the ships and travelled over land as part of their escape plan. Thereafter, as every expedition journeyed north, so more and more of the Canadian north was explored, mapped, and surveyed by those visiting parties.

21. *The Sea of Ice* by the German artist Caspar David Friedrich (1774–1840). Even before the ill-fated Franklin Expedition, European artists and writers were fascinated by sea ice and glaciers. Friedrich's painting depicts a ship being crushed by immense slabs of sea ice.

The Franklin mystery endured and continued to generate huge controversy as Victorian Britain was shocked at the thought of British men resorting to cannibalism. It was not until the 1990s that Canadian archaeologists eventually recovered some skeletal remains of the crew on King William Island. The injuries sustained by the men seemed to bear out claims made by local peoples that the desperate survivors had turned on each other. In 2014, Parks Canada made a monumental discovery. HMS *Erebus* was found in a well-preserved state in the waters off King William Island and two years later HMS *Terror* was found in the appropriately named Terror Bay. In both Britain and Canada, there have been highly popular exhibitions showcasing some of the artefacts recovered from the ships. The discoveries confirmed the veracity of oral testimony by Inuit—had these sources been listened to earlier, the ships might have been discovered rather sooner. The remains of Sir John Franklin have never been found; a heroic statue of him can be seen in central London.

The Northwest Passage was not actually traversed until the early 20th century. The great Norwegian explorer Roald Amundsen travelling in a small fishing boat called *Gjøa* made it all the way through in 1906. While Amundsen later found fame as the first to reach the South Pole in 1911, he achieved something that other European explorers had craved for over 500 years. But Amundsen's voyage took three years at a time when the Panama Canal connection to the Pacific was still under construction and not completed until 1914. Sea ice remained a pressing problem and the promise of open water seemed a chimera. Despite evidence to the contrary, it was still fashionable—even in the early part of the 20th century—to suppose that oceanic currents from the Pacific would warm the colder waters to the north. The Bering Strait does act as a warming funnel helping to reduce and break up sea ice. However, as the US naval expedition led by George De Long discovered in 1879, this warming effect was not sufficient to save the USS *Jeannette*. The ship was caught in the grasp of sea ice and crushed somewhere north of the Bering Strait. The crew escaped the unfortunate vessel and some of the survivors made it across to Russia, where they were rescued. De Long was not amongst them.

While we often think of exploitation in terms of resources such as fish, whales, minerals, and timber in an Arctic context, it is worth bearing in mind that there are other ways in which this part of the world has turned a profit for outsiders. One notable feature of the *Jeannette* expedition was that it was funded by the owner of the *New York Herald*, James Gordon Bennett. In the 19th century 'Arctic fever' was used as a term to describe high-level public interest in the northern reaches of the world. The Arctic was harvestable. Newspaper readers would pay money to read stories about Arctic heroics, painters and writers could sell their Arctic art, and exhibitions could be organized for interested members of the public. German cartographers such as August Petermann (1822–78) bolstered their reputations by promoting his 'open polar sea' theory. Based in London in the 1840s and 1850s, Petermann was a major intellectual influence on British and American Arctic expeditionary planning. He was a highly skilled map-maker, sending maps to the Admiralty in London in the midst of the frantic search for the Franklin expedition. His 1852 map on the 'open polar sea' reinforced this sense that there was a Northwest Passage to be found and exploited. Sadly, Petermann suffered from depression and took his own life in 1878.

Exploiting Arctic lands

The Hudson's Bay Company (HBC) was integral to the early exploitation of the Canadian north. Before Hudson's mysterious death, he worked for the London-based Muscovy Company, which traded with the Grand Duchy of Moscow. The vast Hudson Bay, some 1.2 million km², was the epicentre for his extensive trading empire. Chartered in May 1670, with the patronage of Charles II of England, the HBC specialized in fur trading. Beaver pelts were harvested from the early 16th century alongside the development of the cod fishing industry off the Grand Banks of Newfoundland and the Gulf of St Lawrence. Europeans, especially English and French traders, exchanged metal and cloth goods in return for fur pelts initially hunted by indigenous peoples. Felt proved to be extremely popular and proved ideal material for hats and clothing. English and French traders created strategic alliances with indigenous communities. There were actually multiple 'Beaver Wars'. The HBC's forays further north away from the battlegrounds of the St Lawrence region were an attempt to open up new more peaceful resource fields. In 1670, trading posts began to appear around the coastline of Hudson's Bay. Such was its importance, the beaver featured on Canada's first postage stamp in 1851.

Beaver pelts were exchanged for goods such as blankets, guns, and food supplies. From the 17th to 19th centuries, the trade often led to territorial conflict within indigenous communities as struggles emerged over beaver harvesting. European settlement of the north and west of Canada brought devastating diseases such as smallpox. It was a rapacious trade, but it brought much enhanced income-earning potential for indigenous hunters. Terms such as 'Made Beaver' (akin to currency) were used to help regulate the largely bartering trade with indigenous communities, acting as a unit of exchange value based on one prime male beaver skin. Indigenous hunters were paid with HBC tokens that enabled them to purchase HBC goods and services. Tobacco, cloth, blankets, and guns were popular sales items. It was a cashless circular economy.

As the HBC's trading operations moved westwards so new 'resource booms' emerged. In the 1850s, there were 'gold rushes' as settlers moved into the valleys of British Columbia, as miners from California joined others hoping to find their fortunes. After the Fraser and Cariboo Gold Rushes, the hunt for new resources moved further north. Starting in the 1880s, gold mining in the Yukon reached fever-pitch in the late 1890s when at least 100,000 prospectors travelled north to the Klondike River. Prospectors used wood-fired steam jets to thaw the frozen ground. Towns like Dawson City, which established a gold and pelt trading centre with a population of some 30,000, grew rapidly. By 1900 the boom dissipated with estimates that at least 1 billion US dollars of gold had been extracted during the Klondike rush.

The rush for gold coincided with the peak in beaver pelt trading. By the late 19th century, the Canadian beaver was close to extinction. At the start of the trade in the 17th century, beaver numbers were estimated to be around 6 million. Around 200,000 beaver pelts were being exported annually to European and other foreign markets in the 1880s and 1890s. The stampeders,

22. 'Stampeders' in northern Yukon in 1900 pose with woolly mammoth tusks recovered from the permafrost.

inadvertently, caused a boom in something else. As they dug for gold, they uncovered the remains of ice age animals. Pleistocene palaeontology received an early boost too; miners posed proudly with their ice age finds, including magnificent woolly mammoth tusks (Figure 22).

Changing fashions and growing awareness of the ecological impact of the trade meant that fur hats declined in popularity with European customers by the mid-20th century. But the fur trade did not disappear in the 20th century as new markets appeared in the United States and China. The official seal of the city of New York features two beavers and there was a thriving 'fur district'. Black beaver fur hats were popular with the African American community and considered emblematic of success. In 2019, the city of New York proposed a new fur ban, but this drew some to claim that the ban was an attack on African American culture.

Resource boom and bust cycles were endemic and costly to ecologies and indigenous societies. Even today the fur trade, while controversial, is an important element of the Canadian economy. The main market is Russia and trade continues in beaver, mink, muskrat, and racoon. The industry in Canada employs up to 25,000 indigenous people. Trapping is still important in the sub-Arctic, in areas such as northern Quebec and Alberta, with smaller trapping economies in Yukon and Northwest Territories. The beaver was adopted as the national symbol of Canada in 1975. It is found in every Canadian province and territory.

Mining in the Arctic

There has been industrial-scale mining in the Arctic for at least 300 years, longer if you include operations in southern Finland and Sweden. There is also evidence that coal was used by whalers operating around Spitsbergen (now Svalbard) in the 17th century. The Dutch and English established small settlements on the archipelago and used coal for heating. Coal has been mined in Greenland from the 18th century and cryolite extraction followed in the 19th century. In the Russian Arctic, there has been mining activity from about 1700, initially focused on gold and silver and then expanding into other areas including coal, other precious metals, and diamonds. Metal mining spread to northern Canada and Alaska in the 19th century, as 'gold and silver rushes' encouraged armies of prospectors to seek their fortunes.

In the 19th century, new organizations such as the Geological Survey of Canada (GSC) played an important role in promoting the mining industry in the north. Created in 1842, geologists and geographers of the GSC started to map the country's landscapes and ecologies, identifying areas of mineral potential. Geologists, in particular, were in demand. Mapping the nation's resources was considered by William Edmond Logan, the founder of the GSC, to be integral to a country's commercial potential and future welfare. Logan travelled to the 1851 Crystal Palace Exhibition in London

with samples including bitumen and oil sands to showcase
Canada's mineral wealth.

Others were pushing ahead with the mining of the Arctic. In
contrast to the vast Canadian frontier, Spitsbergen's coal seams
were abundant and accessible and, with no Indigenous and First
Nations peoples to worry about, the island's resources attracted
growing interest from overseas speculators. The main settlement
of Spitsbergen is Longyearbyen and was named after John Munro
Longyear, an American businessman who visited the island in
1901 and created the Arctic Coal Company. Norwegian miners
were hired, and the coal was shipped to European and North
American markets. The Arctic Coal Company did not enjoy a
monopoly. The first commercial mine was actually opened in the
1890s by a Norwegian skipper, Sören Zachariassen. Spitsbergen,
subsequently, experienced a 'coal rush' as others advanced their
claims and opened up mining camps and pits around various
fjords. Arctic mining is a story of repeated booms and busts.
Norwegian, American, and British industrialists transformed
Spitsbergen into a mining hub, and they were joined by Russian,
Swedish, and Dutch parties. Northern Norway was an important
supplier of miners for an industry that was underpinning
European industrial and urban development.

Coal mining in Spitsbergen was not just about servicing the
industrial needs of European economies (Figure 23). Mining
companies became adept at expanding operations so that they
occupied ever more land. The Arctic Coal Company even
demanded the United States claim the islands while British
mining companies such as the Scottish Spitsbergen Syndicate
were adamant that Britain should be a participant in determining
the question of ownership of this no man's land. Conflict was
avoided thanks to a crash in coal prices in the early 1920s and the
negotiation of the Svalbard Treaty, which determined that the
islands should be Norwegian. Importantly, the treaty protects the
rights of other users and ensured that those who wished to mine

23. There is a long tradition of coal mining in Svalbard, as illustrated by this prominent statue of a coal miner in Longyearbyen.

could do so without prejudice. The treaty prohibits military activities.

For the past hundred years, the only mining operations undertaken in Svalbard have been carried out by Norwegian and Soviet/Russian operators. As the strategic significance of Svalbard grew during the Second World War and later the Cold War, neither Norway nor the Soviet Union wished to close their respective coal mines. Given the distance to markets, Svalbard coal has never been cheap. In 2017, after decades of subsidies, the Norwegian government decided to close its commercial mining activities.

The cessation of mining comes at a time when Norway remains worried about Russia's role in the Arctic. A Russian/Soviet community has been active in Barentsburg, the second largest settlement in Svalbard, since the 1930s. Coal extraction continues despite its unprofitability and Russia might decide to open up any mines that Norway selects for closure. The fate of Norway's last coal mine, Gruve 7, will be watched carefully by international observers. At present, the Norwegian mine produces around 100,000 tonnes per year and the Russian operation close to 140,000 tonnes. Norway has been first to blink regarding unprofitability and unsustainability. The state-owned company Store Norske is expected to decrease its operations at the last mine, as pressure mounts to transition towards green energy supplies and address the century-long environmental legacy of mining. From the 1990s onwards, Norway has been encouraging tourism in Svalbard as the main alternative to mining.

Coal mining in Svalbard highlights a major paradox: extracting an expensive fossil fuel for heating and power with the knowledge that the Arctic region is enduring the most severe consequences of global warming. Meanwhile Norway is a major beneficiary of domestic hydropower and a leader in the green transition to electric vehicles.

The Arctic paradox

Arctic observers often use 'Arctic paradox' to describe a series of contradictory pressures facing the region. In 2005, a reporter called Marla Cone wrote a book called *Silent Snow*. Inspired by Rachel Carson, she made the point that Arctic peoples are being contaminated by poisons produced and circulated from elsewhere. Inuit in Canada and Greenland faced a series of health risks as a consequence of their dependency on toxic meat and blubber from seals and whales. But there is another 'Arctic paradox'. Northern peoples depend on resource exploitation including oil and gas (Figure 24) not only to support national economies such as Russia, but also to generate employment and investment opportunities for local communities. People in the north understand that oil, gas, and minerals play their part in climate change and are similarly aware that their own governments have often neglected the social and economic development of Arctic peoples despite all that resource wealth.

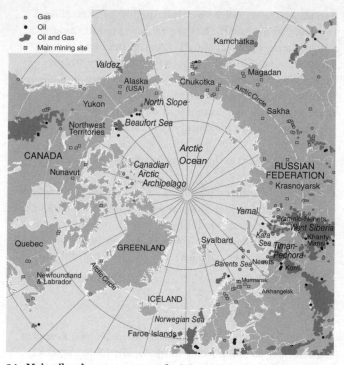

24. Main oil and gas resources and mining sites in the Arctic.

Mining continues to be integral to the history and the future of the Arctic. It has left an indelible mark on Arctic societies and landscapes. For many northern and indigenous communities, the mining sector has been integral to employment, infrastructure, and service provision. The North American Arctic has been a very different region in which to operate compared to the Nordic and Russian Arctic. In Alaska and the Canadian north, remote and smaller communities may welcome mining operations because of those social and economic opportunities. This is often a double-edged affair because any new investment brings with it the potential for conflict with indigenous communities and their land rights. Investment in infrastructure can and does bring with it concerns about ecological impact, pollution, interference with subsistence hunting and herding, and disputes about who should be financing road construction. Even building new roads to connect remoter areas can raise concerns that improved accessibility might facilitate drug smugglers, bootleggers, and criminal networks.

For the residents of Alaska, the discovery and exploitation of oil from the late 1960s brought a yearly dividend paid to all citizens from the Alaska Permanent Fund. The payments started in 1982, after revenue from the oil industry was well and truly established. The dividend varies with oil prices. In a 'good year' an individual might receive around $2,000 with lows recorded of $800. Many Alaskan families depend on their annual payments to meet essential expenditures. A family of four or five could receive anywhere between 8,000 and 10,000 US dollars. Arctic oil is an American experiment in universal basic income.

The settlement of Rankin Inlet (around 2,800 people) in the Canadian territory of Nunavut is supportive of Agnico Eagle's Meliadine gold mine. It started production in May 2019. With an initial investment of around CAN$550 million, the mine will target six gold deposits aiming to extract over 230,000 ounces of gold in 2019. The company estimate that there might be around 3.8 million ounces in reserves. Mining projects are rarely free from complications. Many employees at mines work on a fly in and fly off schedule. They travel into the mining complex from outside the region, work shifts, and then return home. This is not unusual for the mining and extractive sector, but in smaller, remote communities its distorting effects can be keenly felt. Out of town mining staff wishing to move their families closer end up causing housing shortages. Most employees will live their working lives entirely on the mining site.

As a consequence of Canada and Alaska's regulatory environments, mining operators usually contribute to local communities via donations and more substantially to community funds. Since the 1970s, indigenous communities across North America have acquired land and resources rights and 'social licensing' is a major factor in determining whether mining projects go ahead. For an organization such as the Kivalliq Inuit Association, the potential is considerable. This enormous region of Nunavut has around 10,000 residents in an area of over 440,000 km². The Meliadine mine was only possible because the KIA gave its approval for an Inuit Impact Benefit Agreement in July 2015. Agnico Eagle had to undertake to pay royalties and fees while ensuring that opportunities were created for local residents. Agnico Eagle must remain respectful of Inuit values and generate up to 50 per cent of employment for Inuit. In the next ten years, the KIA could have funds running to some £50 million. This investment is being used for education and training as well as provision in areas such as heritage and tourism.

Mining is integral to indigenous futures in the North American Arctic and Greenland. Across the Davis Strait, the government of Greenland has been explicit that mineral resources are key to national self-determination and autonomy. As a former colony of Denmark, Greenland has become ever more autonomous following two landmark referenda in 1979 and 2008. In December 2009, the government in Nuuk gained exclusive rights over sub-surface resources. While Denmark continues to pay a so-called Block Grant worth about £500 million per year and formally address foreign and security policies, local political leaders in Greenland are active players in the global arena. Their message has been uncompromising—'Greenland is open for business.'

This message to the wider world is a complex one. Mining and offshore oil and gas exploitation are seen as key to the country's future and possible independence from Denmark. With a population of just 56,000, there is also recognition that the block grant from Denmark is integral to the economic and social welfare of the country. Fishing and wilderness tourism are major income generators, and this can sit awkwardly with the promotion of mining. While mining is an established industry on the island, newer projects addressing uranium, diamonds, rubies, and zinc have proved controversial. Most of the mining in Greenland occurs in the south and west of the island, with global interest most acute in access to rare earth metals. Lithium and cobalt, for example, are crucial to green technologies, smart phone batteries, electric cars, and military applications. Recognition of Greenland's strategic importance and mineral potential might explain President Donald Trump's reported comments in the summer of 2019 that he was interested in purchasing the world's largest island. His offer was quickly rebuffed by the governments of Denmark and Greenland.

Chinese investment in mining and infrastructural projects in Greenland has been cited as a source of strategic concern for the USA, all of which has encouraged new interest from the United States in investing more in aerial surveys and mining agreements with Greenland. The US Geological Survey will be leading on a survey of the south-western province of Gardar to help inform both strategic decision-making of the government of Greenland and future regulation of resource extraction. NASA's annual Operation IceBridge, using ice-penetrating radar, has assembled a remarkable three-dimensional map of the Greenland ice sheet. What is striking is how the same platforms and radar technologies are being used for very different purposes: assessment of mining potential on the one hand and glaciological monitoring on the other.

Complicating all of this is that the USA has a long-standing relationship in Greenland with its NATO ally Denmark. As a consequence of Cold War tension, the USA established a military base at Thule in the far north of the country designed to intercept Soviet bombers flying over the Arctic Ocean. Mining was not integral to that strategic calculation. Seventy years later, Greenland is now a resource hotspot and an epicentre of concern about climate and sea level change. An increasingly independent government in Nuuk is quite prepared to negotiate with international actors including Australia, China, the UK, and others in the pursuit of financial independence and autonomy. Onshore and offshore, Greenland's resources are being mapped, surveyed, and licensed for potential extraction. Prospects for an independent Greenland will rest on future revenue streams. In April 2020, President Trump offered Greenland a $12 million aid package and announced that the United States would open a consulate in Nuuk.

Legacies of resource development

Resource development in the Arctic has bequeathed some intriguing legacies for the countries and communities implicated in it. For example, iron ore mining in northern Sweden has been so extensive in terms of excavation that the city of Kiruna has been literally undermined. While communities have lived in the north for thousands of years, the establishment of Kiruna in 1900 owes a great deal to the discovery of iron ore in Kiirunavaara Mountain. First surveyed in the 18th century, significant iron ore extraction did not really develop until the late 19th century. Thereafter, the mining town grew to around 20,000 as the state-owned company, LKAB (in Swedish, Luossavaara-Kiirunavaara Aktiebolag), began extraction on an industrial scale.

In the post-war period, Sweden remained an active mining economy supplying 90 per cent of Europe's iron and holding at least 5 per cent of the world's reserves. The mine in Kiruna remains the most prominent supplier. Because of the vast mining operation, the town has had to take the momentous decision to relocate at a cost of at least £750 million. The target date for completion is 2035. Such is the profitability of the mine, there is no interest in closing it.

Moving an entire town because of mining is one notable legacy but there are others. In the former Soviet Union, hundreds of special towns were established as the north of the country was targeted for industrial development. Initiated in the 1920s and 1930s, populations were moved to create these new settlements, some of which were 'closed cities', where few could visit. The forced labour of political prisoners under the Stalin era constructed canals, roads, railways, and even new resource cities such as Norilsk, which was for the specific purpose of mining nickel. Hundreds of thousands perished in horrific working conditions with winter temperatures plunging to –50°C. Those who survived were often prevented from leaving.

While most of the 'closed cities' were directly related to nuclear weapons development or centres for military operations, others were single-industry towns. In the Soviet Far East, the city of Kadykchan was established during the Second World War for the sole purpose of coal extraction. For decades, miners worked two coal mines called Number 7 and 10. For those who were not political prisoners, moving to a resource town was not all bad news. For skilled miners and loyal communist party members, the quality of housing provision and salaries were good by Soviet standards.

In more recent times single-industry cities across the Russian north have been hit by population decline and out-migration, declining profitability of markets, and concerns about the toxic legacies of dust and 'black snow' laced with traces of mercury and arsenic. Local communities have taken to social media to photograph and comment on the highly polluted areas of the Russian Far East. Families are particularly concerned about the health of their children and accuse the local and regional authorities of being unwilling to confront the unfolding health and environmental crisis. With some 2 million people living in the Russian Arctic, the resource sector has been the biggest employer and many towns have been single industry since the 1930s and 1940s. Economic and social diversification is especially challenging when heavily polluted towns are not likely to attract incomers.

The Russian leadership is clear that the Russian north is a vast resource base, which will underpin future options in the 2020s and 2030s. Natural resource development and maritime transport are strategic priorities. All of this goes hand in hand with military and civilian infrastructure investment. Russia's new generation of icebreakers (Project 10510 and Project 22220) is formidable. The Leader class is around 70,000 tonnes and many times bigger than US icebreakers such as the *Polar Star* at 10,000 tonnes. The Russian vessels will be nuclear powered and even the smaller Arktika-class ships will be around 30,000 tonnes.

Resource development in the north continues to be identified as integral to the existence of Russia. Climate change in the Arctic has not dampened interest in hydrocarbon extraction. If anything, perceptions of a 'melting and thawing' Arctic have encouraged further Russian investment in energy, transport, and defence. President Putin supported the Paris Climate Accord in 2015 even if he fully appreciates his country's dependence on oil and gas extraction. The Russian north faces very real problems. The structural stability of natural gas and oil infrastructure could be compromised by permafrost thawing. When permafrost thaws, the bearing capacity of the ground disintegrates. Expensive infrastructure can buckle and collapse leading to leaks, expensive repairs, and delays to production. The Russian Arctic is a high-cost production environment, so delays are costly. Newer projects such as Bovanenkovo and Yamal Liquid Natural Gas (LNG) plants have used reinforced foundations to cope with permafrost-related instability. Old infrastructure at established natural gas fields is vulnerable to disruption.

The Yamal Peninsula is at the heart of contemporary Russian energy development, with Star Wars-like infrastructure facilitating extraction and transportation (Figure 25). Around 20 per cent of Russia's GDP comes from the north and LNG development is pivotal to Putin's vision for the Russian Arctic. The Russian energy operator Gazprom (a major sponsor of Champion's League Football in Europe) estimates the reserves of the Peninsula alone are at least 55 trillion m^3 of gas with possibly 40 trillion m^3 offshore and in adjacent fields. Russia hopes to supply East Asia with oil, gas, and other minerals.

As some parts of the Russian Arctic are depopulating and condemned as highly toxified, new investment is flowing towards offshore gas development. Strategic relationships with China and others such as Vietnam are playing their part in creating new energy partnerships. The Northern Sea Route (NSR) that links ports and cities around the north of Russia is fast developing as a global transit space. Russia's economic development goes hand in hand with a determination to ensure that its northern ports such as Murmansk and Arkhangelsk are kept secure. The country's Northern Fleet and nuclear deterrent are stationed in the western portion of the Russian Arctic. Economic development in the form of mining and resource extraction are existential matters in Russia. They speak directly to the future survival of the Russian Federation and its ability to manage its own resources (Figure 25).

Indigenous peoples and resources

For indigenous peoples throughout the Arctic, the acquisition of land and resource rights has meant that contemporary resource development must be a great deal more attentive to their wishes and interests. Mining and energy industries provide invaluable

25. The Gazprom Arctic Gates oil loading terminal on the Yamal Peninsula in Siberia.

revenue streams that actively contribute to investment in community infrastructure and cultural and social heritage. Millions of dollars, roubles, and krone will make a meaningful difference to communities often neglected and marginalized in the past. On the other hand, resource development can be controversial and disruptive to those same communities. Traditional lifestyles such as reindeer herding can be heavily interfered with as energy and transport infrastructure disrupt migratory flows of reindeer and caribou.

Energy projects attract interest from outside parties. Environmental groups are critical of mining ventures and throughout the Arctic 'green' activists are often mistrusted and disliked. It is not uncommon to hear northern communities complain about 'green colonialism', where any form of resource exploitation in the Arctic is questioned and criticized by outsiders. Whether it is mining or harvesting marine mammals, Arctic economies are for the large part grounded in resource development, some of which will be driven by mining companies and national governments eager to export to global markets. Living and working sustainably in the Arctic requires trade-offs. Mining provides job opportunities. Hunting seal, whale, polar bear, and moose offers food security. Selling fur pelts provides another revenue stream to Arctic communities. External campaigning designed to 'Save the Arctic' can end up perpetuating old-fashioned stereotypes about Arctic peoples and landscapes.

Criticism of northern communities and their resource usage sits uneasily with an Arctic that is being scrambled by climate change. A warming Arctic is disrupting traditional fishing activities, with warmer waters either proving harmful for native species such as salmon and/or disruptive to traditional hunting seasons. When sea ice is thinner and fragmented it becomes more dangerous to hunt marine mammals at the ice edge and more open water allows winter storms to cause more damage to shorelines. Thawing permafrost contributes further to this sense of discombobulation as the 'natural assets' of the Arctic literally give way. They might not be 'resources' to casual observers but northern communities are carefully calibrated to the changing rhythms of light, darkness, warmth, and cold.

In recent seasons, fishing companies have noted that their quotas are being affected by warming waters. Fish stocks such as Arctic cod face severe depletion, and possible collapse in some parts of the Barents Sea. Diminishing sea ice cover does not automatically confer resource-based advantages even if new fish species are entering the Arctic. Crab and fish harvesting have been adversely affected by migratory change and poor sea ice conditions. Longer and more intense wildfires and abnormal summer temperatures threaten fishing potential in rivers and lakes. Environmental change in the Arctic is complicated and any ecological niche advantages might end up being accrued by alien species.

Resource futures

In October 2019, the Sámi Arctic Strategy was launched at the
EU Arctic Forum and it threw down the gauntlet to national
governments in Norway, Sweden, and Finland. The strategy
demanded that Sámi pursue their vision for the Arctic based on
human security and sustainable stewardship of the land, water,
and biological resources including reindeer. Indigenous
knowledge was pivotal to this vision, as was a collective desire to
avoid a past history of marginalization and subjugation.

Arctic warming brings costs and opportunities for indigenous and
settler communities. Commercial and traditional fishing might
benefit from southern species such as pollock, mackerel, and
Pacific cod moving northwards towards the Arctic Ocean. In 2018
the Agreement to prevent unregulated high seas fisheries in the
Central Arctic Ocean was signed by ten parties including Canada,
the United States, Norway, Iceland, and China. Indigenous
peoples were represented on several national delegations. The
Agreement calls for 'local and indigenous knowledges' about the
Arctic Ocean to be respected. While it imposed a moratorium on
fishing in the Central Arctic Ocean, it did acknowledge that
commercial fishing could become a reality in the future. Even in
the 1970s, it would have been unthinkable that we might witness
fishing boats operating in the international waters of the high
north or what used to be known as the open polar sea.

The resource potential of the Arctic continues to evolve, and
indigenous and northern communities hope they will be
beneficiaries in any future resource booms. States are also going to
be wedded to resource strategies. It is difficult to imagine Finland
not harvesting timber for many years to come. For centuries,
Finland supplied the European shipbuilding industry and enabled
the imperial fleets to develop in the pre-steel era. In the 17th
century, Finland was the largest producer of pine tar, which was
used in ship hull construction.

Finnish Sámi have been at the forefront of clashes over the impact of timber on traditional reindeer herding. The state timber agency in Finland must negotiate every year with Sámi with regard to proposed tree-felling. But timber is still big business in Finland, including Finnish Lapland. Pulp, paper, and timber are major exports to the rest of the European Union. The Finland landscape is 75 per cent woodland and timber exports represent around 8 per cent of GDP (2018). Biomass is also used for power generation. Every resource can be used to tell a story about how the Arctic has been exploited and developed.

Chapter 6
Arctic governance

The Arctic, unlike many other parts of the world, has been spared military conflict, civil wars, and terrorism. While the Second World War did affect high latitude locations such as Alaska, Svalbard, and the maritime convoy routes in and out of Russia, the post-1945 era has been a curious mixture of Cold War antagonism coupled with a history of circumpolar cooperation. In 1973, for example, five Arctic countries including the United States and Soviet Union signed the Agreement on the Conservation of Polar Bears at the same time the superpowers were busy navigating their nuclear-powered submarines under the Arctic sea ice. The Arctic remains heavily militarized in the sense that military exercises and patrolling are the norm; yet all Arctic states work together in areas such as search and rescue, environmental protection, information-sharing, and scientific cooperation.

Arctic governance is a complex field. For those living outside the Arctic, it is tempting to think that the absence of conflict is due to the mitigating role of environmental factors such as the long polar night, extreme cold, remoteness, limited infrastructure, and sparse populations. While these are all important, Arctic communities vary in size from small villages to cosmopolitan cities such as Anchorage, with a population over 290,000 people. The largest Arctic cities such as Anchorage and Murmansk (population around 300,000) are similar in size to Iceland (350,000) and much larger than Greenland (56,000). Infrastructure provision is dense in the European Arctic but spartan in Greenland and northern Canada. Established activities such as mining and fishing mean that parts of the Arctic are highly industrialized, while others lack access to basic amenities such as safe drinking water and adequate housing. Climate change in the Arctic is magnifying these pressures; permafrost thaw, sea ice loss, and wildfires all place further pressures on existing infrastructure and generate security fears that the region is changing too rapidly.

While the Arctic is vast, and populated by only 4 million people, it has, for the past three decades, stimulated a shared culture of international cooperation. Governance in the Arctic is highly diverse, depending on the level of local autonomy and the place of indigenous people within dominant political-economic systems. Greenland is the most notable example of indigenous self-governance. By way of contrast, indigenous peoples in Russia are very much 'junior partners' in the governance of the Federation's northern territories. Russia is extremely sensitive to 'foreign influences', and indigenous organizations such as RAIPON have fallen foul of accusations that they are working against the interests of Russia.

What do we mean by 'Arctic governance'?

Arctic governance, like governance more generally, involves an array of actors, legal regimes, institutional and social contexts, and strategic aspirations such as economic development, regional autonomy, environmental sustainability, and national security. As one moves across the North American, European, and Russian Arctic(s), the meaning and relevance of 'Arctic governance' changes. In the far north, the international waters of the Central Arctic Ocean (CAO) are now attracting far greater attention, as the ongoing effects of climate change 'open up' the most northern territories of the world to global interest and scrutiny.

None of this means that the Arctic is short of governance. One of the great myths perpetuated about remote places is that they exist in a vacuum. The Arctic is not the 'wild west'. The maritime Arctic, including the CAO, is governed by an international legal framework called the United Nations Law of the Sea Convention (UNCLOS). In 1982 UNCLOS established rules for the seabed, the water column, and the surface for every sea and ocean in the world. The Arctic Ocean may be the smallest of the world's oceans, but it is still larger than Siberia in area and, in places, extends to depths beyond 5,500 m.

On land, the Arctic falls under the sovereign authority of the eight Arctic states and their indigenous peoples and northern communities. One of the most notable trends in Arctic governance is the shift, since the 1970s, towards land claim agreements, devolution, and expressions of autonomy. Indigenous peoples in Alaska, northern Canada, and Greenland have secured substantial new rights to land including sub-surface rights to minerals and other resources. While the status of indigenous peoples around the world varies enormously, Arctic governance and indigenous rights are closely entwined. Mapping of Arctic territories, landscapes, and resources is a key part of these processes and not immune to high profile provocation. The distinguished oceanographer and Hero of the Soviet Union, Artur Chilingarov, was instrumental in planting a Russian flag on the bottom of the Arctic Ocean seabed in August 2007, descending some 4000 m (around 13,000 feet) in a submersible to do so.

The Arctic Council and circumpolar cooperation

In the dying days of the Cold War, Arctic parties turned their attention to new relationships with one another. Inspired by the geopolitical thaw between the Soviet Union and the United States, environmental and scientific cooperation was widely regarded as a mechanism for building goodwill. As the Cold War demonstrated, scientific cooperation flourished at particular moments, especially in the early 1970s when US–Soviet scientific exchanges were commonplace. The renewed tension of the late 1970s and early 1980s, however, saw the exchange and workshop programme decline markedly. With détente returning in the late 1980s, Finland emerged as a leading proponent of a new initiative.

In 1989, Finland approached the other seven Arctic states with a proposal for what became known as the Rovaniemi Meeting, for the northern Finnish city in which it was held. At the meeting, the Finnish government chaired a discussion about the protection of the Arctic environment. What emerged was an agreement in 1991 to develop an Arctic Environment Protection Strategy (AEPS). The AEPS identified areas of common interest ranging from environmental monitoring to emergency preparedness and ecological protection. Without pointing the finger at any one party, the Arctic states worked collaboratively. Significantly, the negotiation of the AEPS involved three indigenous peoples' organizations—the Sámi Council, the Inuit Circumpolar Conference (later Council), and the Association of Indigenous Minorities of the North, Siberia and Far East of the Russian Federation. The AEPS was observed by other European states including Germany, Poland, and the UK.

The AEPS declared that all parties were committed to the environmental protection of the Arctic while being mindful of the 'needs and traditions of Arctic native peoples' and the need to eliminate pollution from the Arctic environment. At the time of the final agreement in June 1991, it was abundantly clear that the Arctic was not only going to have to address the toxic legacies of Cold War militarism (such as abandoned nuclear submarines, decaying military surveillance infrastructure, and discarded waste), but also long-range pollution from elsewhere. Persistent organic pollutants (POPs) such as polychlorinated biphenyls were identified as particularly troublesome. Once in the Arctic, the POPs were 'trapped' due to the cold and infiltrated food webs to contaminate seals and whales via bioaccumulation. Indigenous peoples eating marine foodstuffs were particularly vulnerable. There was a particular worry that pregnant women in northern communities might be passing on toxins to the next generation. The parties to the AEPS agreed that this problem needed to be addressed urgently and committed them to an Arctic Monitoring and Assessment Programme (AMAP) designed to monitor POPs in air, water, and biota.

Serious as this problem was to the Arctic, it helped to galvanize the Arctic states and their indigenous peoples. The enduring presence of far-travelled pollutants, including atmospheric pollution known as Arctic haze, served as a timely reminder that the ending of Cold War hostilities provided an opportunity to build governance structures and tackle common environmental concerns. In 2020 microplastic fibres were detected across the Arctic Ocean and sourced to urban wastewater from rivers in Europe and North America. The AEPS provided an important stepping-stone for a scaling up of circumpolar ambition. In 1993 the parties decided that the promotion of sustainable development and environmental protection should be at the heart of their collective ambitions. Three years later, the Canadian government brought all parties together to establish an intergovernmental forum called the Arctic Council. The 1996 Ottawa Declaration stated that it would:

> provide a means for promoting cooperation, coordination and interaction among the Arctic States, with the involvement of the Arctic indigenous communities and other Arctic inhabitants on common arctic issues, in particular issues of sustainable development and environmental protection in the Arctic.

The Ottawa Declaration is not a treaty. As with the AEPS, it lacks the legal authority of a treaty or convention. It does not speak of financial and legal responsibilities and, in order to secure consensus, the eight parties had to agree that it would not consider military security. The focus of the nascent Arctic Council, as with the

AEPS, remains sustainable development and environmental protection. Respectful of the sovereign rights of the eight Arctic states, the Arctic Council concentrates on circumpolar cooperation and information-sharing. Notably, however, it acknowledges Arctic indigenous peoples as permanent participants. As the declaration notes, 'The category of Permanent Participation is created to provide for active participation and full consultation with the Arctic indigenous representatives within the Arctic Council.'

The creation of the Permanent Participant (PP) category was explicit recognition of the interests and wishes of the Arctic's indigenous peoples. In the earliest meetings of the Arctic Council, a pattern was established that PP representatives would sit around the same table as Arctic state leaders and officials. To tackle both long-term and time-sensitive topics, the Council established six working groups such as the Sustainable Development Working Group and the Protection of the Arctic Marine Environment to work alongside task forces for Marine Cooperation and Improved Connectivity in the Arctic.

The Arctic Council allows for external observers to attend ministerial meetings and participate in working group and task force activities. These simply observe the business of the Arctic Council but do not decide the priorities of working groups and task forces. In May 2013, the Arctic Council approved observer status to China, India, Singapore, Japan, and South Korea. In 2017, Switzerland, the West Nordic Council, and the World Meteorological Organization also joined as observers

The most important observer to join the Arctic Council in recent years has been China. All observers must respect the sovereign rights of the eight Arctic states and agree to strict rules of engagement. Observer requests are considered every two years at the Arctic Council ministerial meeting and any approval has to be unanimous. Estonia formally applied in 2020 and in early 2021 the EU had still not been approved. Others such as Ireland, Turkey, and the Czech Republic want to become observers in the future. The Arctic Council operates a chairmanship model, and every Arctic state takes it in turn to be chair. Every two years, observer requests are considered at the ministerial meetings attached to the Arctic Council. Canada and Russia have expressed concern that there are nearly forty observers and have argued this number needs to be capped.

Since its formation, the Arctic Council has established itself as the most notable intergovernmental forum for the Arctic. Despite geopolitical crises such as Crimea/Ukraine, there has been no suggestion that Russia would leave the Arctic Council. The United States has never disengaged from the Council even when the Trump administration broke from the 2015 Paris Agreement. PPs are formally acknowledged as equal partners and the growing number of international observers demonstrates that the legitimacy of this forum is widely acknowledged. Under the auspices of the forum, the Arctic states have signed legally binding agreements on search and rescue cooperation (2011), the prevention of oil spill pollution (2013), and scientific cooperation (2017).

However, we need to qualify the claims we make about the efficacy of the Arctic Council. In any reasonable audit, we would note positives such as cultivating a mechanism for peaceful coexistence and formal recognition of indigenous peoples. We would point to the Arctic Council's scientific reports and publications, which have been hugely influential in shaping global environmental debates. The 2005 *Arctic Climate Impact Assessment* is a stand-out example that brought to the fore the impact of ongoing climate change on Arctic ecologies and cultures. On the negative side, the Arctic Council has to operate by consensus and the eight states are not just Arctic states per se. Some Arctic states such as Iceland and Norway are eager to see NATO play a more active role in the Arctic, while fellow NATO member Canada is more cautious because it does not want to see further militarization. Russia and the United States are challenging partners; bending the norms and rules of international behaviour has defined both Putin and Trump. At the 2019 Arctic Council Ministerial Meeting, US Secretary of State Mike Pompeo was explicit in his criticism of other parties, including Canada, Russia, and China. It was seen as an unhelpful and bombastic speech by many observers. The 2021 Arctic Council Ministerial in Iceland was a great deal more convivial by contrast.

China is a key player to watch now and in the future. It has declared itself to be a 'near-Arctic state' and has plans to establish a 'polar silk road', to build closer economic links with Russia and Nordic countries. Arctic tourism is a major income earner for northern towns and regions such as Kirkenes and Rovaniemi and the Chinese market is strong, with growing interest in Arctic Russia (Figure 26).

26.   A party of Chinese tourists in Murmansk in January 2020. In the background is the nuclear-powered icebreaker *Lenin*, which is now a museum and part of a tour of the port area. Recent years have seen a rapid increase in the number of tourists from China to the Russian Arctic although the COVID-19 pandemic saw a marked fall.

Russia and China are working together to develop natural gas extraction in the Russian north. China participated in a new agreement in 2018 to prevent unregulated fishing in the Central Arctic Ocean (CAO). China's economic size and projection of power capabilities pose real challenges. Industrial pollution from north-east China is a serious problem for Arctic environments. China's fishing fleet and icebreakers such as the *Snow Dragon* (*Xuě Lóng*) are likely to become more active in the Arctic Ocean. There is concern that if fishing develops in the international waters of the CAO, China will do little to discourage illegal fishing. Western observers are worried that, in the worst-case scenario, the CAO will become a hotspot for geopolitical tension with Chinese fishing vessels acting as what has been described as a 'maritime militia'.

Another notable actor is the European Union. Notwithstanding its inability to secure a permanent observer position to the Arctic Council, three EU member states (Denmark, Sweden, and Finland) are integral to the European Arctic. The EU's Northern Dimension policies in the 1990s were explicitly predicated on the fact that new Nordic membership (e.g. Finland in 1995) meant the EU extended all the way to the Russian border. Other parties such as Norway and Iceland are closely aligned with Brussels. The EU is a major funder of Arctic science; it is party to the Central Arctic Ocean fisheries agreement and is a major investor in northern infrastructure and trade. Since 2008, the EU has been refining its Arctic policy and iterations in the 2020s are likely to emphasize the EU's continued role in climate change, sustainable development in the European/EU Arctic (including associate members such as Iceland and Norway), and international cooperation across the region. The international waters of the Central Arctic Ocean, the resources beneath the seabed that Artur Chilingarov visited, will remain part of the EU's policy portfolio.

When the EU announced in December 2019 that it would update its Arctic vision, its statement epitomized the framing of what we would term a 'global Arctic':

While recognising the primary responsibility of the Arctic states for the development of the region, the [European] Council notes that many of the issues affecting the region are of a global nature and are more effectively addressed through regional or multilateral cooperation, in particular, the Arctic Council [as well as] the UN system.

Because of its economic and political clout, EU policies and investment decisions inevitably shape the governance of the Arctic.

Future governance of the Arctic

The eight Arctic states will remain significant players in the future governance of the northern latitudes, but they will not be alone. And there will be some powerful drivers that ensure that the 'global Arctic' as opposed to the 'circumpolar Arctic' will be prominent in multiple ways. The five most important drivers are ongoing climate change, the return of great power competition, indigenous autonomy, technology, and international trade.

First, the Arctic will continue to warm, melt, and thaw. The 2019 Arctic Report Card from the National Oceanic and Atmospheric Administration, a US federal agency, makes for salutary reading. It is clear about the geophysical and ecological changes affecting the Arctic. Why does this impact upon governance? Migratory patterns of marine species provide a good example. As the Arctic Ocean loses sea ice and absorbs more heat, so the living conditions alter. We are already witnessing Atlantic and Pacific fish stocks moving northwards. New areas of the maritime Arctic, such as the Central Arctic Ocean, are receiving unprecedented attention. For the first time we now have a moratorium on commercial fishing in a part of the Arctic where, in the recent past, it would have been inconceivable that commercial fishing would even be possible. Fishing is not happening at present around the North Pole, but it might do in the future.

Climate change brings both new opportunities and costs to those seeking to govern northern territories. For Russia, the sea ice decline might appear to be a good thing because it makes their vast north more accessible to shipping (Figure 27). The strategically significant Northern Sea Route, however, demands more investment for its management and control. As the largest Arctic state, anything that Russia does in the region will appear out of proportion to other Arctic states. What alarms international observers, however, is Russia's concerted efforts to exert ever more control on those who wish to cross the Northern Sea Route.

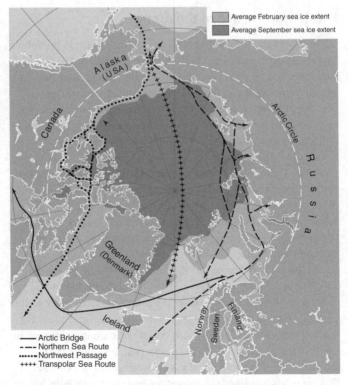

27. Arctic shipping routes and average sea ice extent in the satellite era.

Appeals to security, stewardship, and sovereignty are used by many states, and not just in the Arctic, to restrict the freedom of others. Russia was quick to proclaim the importance of the Northern Sea Route when the Suez Canal was briefly blocked by a giant container ship in March 2021. Thawing permafrost not only releases carbon dioxide and methane into the atmosphere, but also imposes direct costs on the communities and ecologies affected. Infrastructure such as bridges, buildings, and pipelines is already becoming more expensive to maintain. And costs are already high in the Arctic. This does not mean that Russia is going to 'lose its territories' but being an Arctic state is going to become more expensive.

Recent US administrations, while they have differed about the urgency of climate change, share a common concern about the changing nature of the Arctic. The US Air Force, Army, Coastguard, and Navy accept the reality of Arctic transformation and acknowledge new strategic challenges posed by sea ice loss. Alaska faces profound challenges as coastal communities face ever greater problems due to sea level rise and the cumulative damage done by winter storms no longer buffered by sea ice. Permafrost thawing has been found to undermine military runways, hangars, and outbuildings.

Second, it is very likely that great power competition will return to the Arctic. It will be very different from the Cold War antagonisms in large part because China is the new geostrategic player. Ever since the 2014 Crimea crisis and the imposition of sanctions by the EU and USA, China and Russia have sought a closer partnership. The Arctic has been portrayed as a low-tension area since the early 1990s. This is changing: NATO partners identify Russia as a 'persistent threat' and China as a 'challenge' and the northern flank (Arctic and North Atlantic) has been identified as an area of considerable strategic concern. The Chinese influence in Russia, through strategic investment in energy development and broader plans for a so-called 'polar silk road', concerns the

United States deeply. Yet areas of circumpolar cooperation endure, such as the Arctic Coast Guard Forum, which involves all eight Arctic states meeting annually to foster environmentally responsible and safe maritime activity.

The Indo-Russian relationship might be another dimension to watch in the future. In September 2019, Prime Minister Narendra Modi visited Russia and agreements were reached on natural gas, shipping, and defence as well as trade. In the new trade deal, it was agreed that India would extend no less than $1 billion worth of credit for economic development in the Russian Far East. The deal matters because this vast region is suffering from long-term depopulation, environmental disruption, and spartan infrastructure. As a resource economy, Russia will need to continue to exploit the resources of Siberia and other eastern territories. It matters in security terms because underdevelopment and population loss make Russia vulnerable to separatist pressures and, paradoxically, mounting Chinese investment and influence, including in-migration. Both India and Russia have common cause to temper mounting Chinese investment and strategic power while remaining eager to take advantage of business opportunities.

Third, the Arctic will continue to change for indigenous peoples. While there is plenty of work to do in improving the living conditions and prospects for many communities around the Arctic, it is quite possible that we might witness the emergence of an independent Greenland sometime this century. In their latest intelligence assessment (2019), the Danish government has identified Greenland as their number one priority. Its growing autonomy and desire to develop has allowed others—such as Chinese and Australian mining, shipping, and infrastructure companies—to explore investment opportunities. The Danish government has been forced to intervene in areas such as proposed Chinese investment in Greenlandic airports because of security concerns. This is not going to diminish. An independent

Greenland might be enabled by overseas investment and, on securing independence, it might decide in due course to leave the NATO alliance and ask the USA to remove its strategic presence at the northern base of Thule.

Indigenous autonomy is also being empowered by international frameworks such as the UN Declaration on the Rights of Indigenous Peoples and national agreements on land claims and sub-surface rights. From native corporations in Alaska to indigenous groups in Scandinavia, new networks and relationships are emerging in the Arctic. South Korea is developing an interesting partnership with the Aleut International Association via the Korea Maritime Institute's Arctic Academy. Every year indigenous Arctic students travel to South Korea to learn more about that country and the KMI has invested in a community-led indigenous mapping project in Alaska. The distinction between Arctic state, permanent participant, and non-Arctic state, as sanctified by the Arctic Council, will become increasingly blurred. Indigenous Arctic groups can use overseas partners to leverage more resources and political attention from their home states.

Fourth, technology will continue to be a game changer for Arctic governance. New ship designs and innovation in autonomous capabilities are likely to contribute to a further 'opening up' of the maritime Arctic for energy extraction and commercial transportation. Better GPS technologies will be integral to the promotion of a 'safe Arctic'. A melting Arctic is not axiomatically a less hazardous one. Improving satellite connectivity will not only radically change the operating environment for business and the armed forces of the Arctic states, but also for Arctic communities who suffer from low quality yet expensive internet access.

The Arctic has already proven a popular place for the hosting of data servers. Facebook has established a new site close to the Arctic Circle in Sweden (Figure 28). In September 2016, Facebook CEO Mark Zuckerberg visited the site and posted photographs

28. Facebook's data storage centre in Luleå, Sweden, close to the Arctic Circle. It handles all data processing from Europe, the Middle East, and Africa and is the most northerly server hub of this size.

and commentary on his personal Facebook page. As he noted thereafter:

> We're starting deep in the forests of northern Sweden with the Luleå data center. It is a key part of our global infrastructure, and it uses a variety of local natural resources to increase efficiency and save power. The small town of Luleå is less than 70 miles south of the Arctic Circle, and it's typically pretty cold.

Data centres need plenty of energy and they need to be kept cool—something that might become ever more challenging if the Arctic ceases to be reliably 'pretty cold'. Facebook is not alone in investing and innovating. In 2017, the Chinese company Huawei was responsible for establishing the 4G network in the Faroe Islands, which are part of the Kingdom of Denmark, with discussions for a 5G upgrade, although parties such as the USA have warned Denmark and the UK of the potential for exposure to critical infrastructural interference. This will be a recurring theme in other areas such as plans for Russian, Finnish, and American underwater communication cables in the Arctic Ocean. As with the quest for the Northwest Passage, the Arctic offers a shortcut for cable connections between Europe, North America, and Asia.

There is also huge investment in a new generation of oil and gas technologies addressing the technical challenges of working in the Russian Arctic. Underwater drone technology is used to assess operating conditions. More seriously, in December 2019, it was revealed that US technology in the form of gas power turbines was going to be secretly exported for use in the Russian natural gas sector in violation of US sanctions against Putin's Russia. Calls for a high-tech 'smart Arctic' come with governance challenges; greater access and improved safety carries with it benefits for mining and hydrocarbon sectors as well as high-tech enterprises.

Finally, Arctic governance is going to be shaped ever more by trade. An ice-free Arctic Ocean in the 2040s and 2050s will see more trans-polar shipping, a prospect noted by the Arctic Council's Arctic Marine Shipping Assessment (2009). This will mean new opportunities for gateway states such as Iceland, the UK, and Norway to support Asian states and companies eager to steer a course across the North Pole to avoid Russian control of the Northern Sea Route. The trans-polar route would largely involve the international waters of the Central Arctic Ocean and thus evade national controls. More shipping means more pollution of Arctic waters. China has spoken of this trans-polar option as the 'Central Passage' and it is likely that we will see Chinese ships making those voyages as soon as is feasible in summer months. A second Chinese-built icebreaker, *Snow Dragon 2*, is now operational. The Arctic is likely to see more nuclear-powered icebreakers and submarines, as well as Chinese tourists and entrepreneurs who might one day land at Arctic seaports and airports funded by Chinese enterprises and investment banks.

A future Arctic promises to be radically different from the one imagined in the 1990s where eight Arctic states developed a vision of circumpolar cooperation with limited outside interference. NATO, China, the European Union, India, and others all recognize the Arctic as strategically important. While conflict may be a remote possibility, it is not inconceivable that the northern waters of the planet will witness more military activity as rival naval powers seek to enforce their interests. Globally, there is widespread recognition that ongoing climate change is going to make itself felt on the Arctic's sea ice and permafrost. It is difficult to imagine the 'global Arctic' becoming any less important during the rest of the 21st century.

Chapter 7
The Arctic carbon vault

A disproportionately large share of Earth's organic carbon is sequestered in the Arctic, both in the frozen ground and within the shelf sea sediments of the Arctic Ocean. Arctic soils account for only about 15 per cent of soil area on Earth, yet the organic carbon stock in the upper 3 m of the permafrost is roughly equivalent to *half* the total soil carbon, to this depth, in the rest of the world. Soil carbon is especially important because, globally, the amount of carbon in soils far outstrips the carbon storage of all biomass *above* ground. It would be difficult to overstate the significance of the frozen carbon locked in the permafrost. It currently stores about 1,600 billion tonnes of carbon, which is more than double the carbon stock of the global atmosphere and four times the amount of carbon added to the atmosphere in the industrial era. The Arctic provides a planetary-scale ecosystem service by keeping a major component of the Earth's carbon budget in long-term storage.

In striking contrast to the equatorial rainforests, where the great bulk of the biomass is above ground and rapidly decomposed and recycled, most of the organic matter in the tundra ecosystems of the Arctic is *below* ground, within the frozen earth. Permafrost soils or *gelisols* typically have organic carbon concentrations of between 2 and 5 per cent which is roughly 10 to 30 times the amount of carbon found in most deep mineral soils beyond the permafrost zone. Where has all this carbon come from and why is the Arctic such an effective sink?

Plants take in carbon from the atmosphere as carbon dioxide (CO_2). During photosynthesis, carbon molecules are used to create sugars, proteins, and lipids for plant growth. While plants grow only very slowly in the short growing season, in a cold Arctic the decomposition of organic carbon proceeds even more slowly, leaving an organic matter surplus that can be stowed away in the permafrost. Over time, this organic carbon enters deep storage via a range of processes including cryoturbation (the churning of deposits during freeze–thaw cycles), and burial by windblown silts (known as loess) or by layers of alluvium on floodplains and deltas during spring melt floods. Much of this buried soil carbon lies dormant in the Arctic deep freeze where it cannot easily be decomposed because microbial activity is effectively shut down. This long-term excess of organic matter accumulation over decomposition has created a massive reservoir of below-ground carbon. During the summer thaw, waterlogging encourages the formation of peat and peat-rich soils. Almost 50 per cent of the world's peatland carbon is found in the high latitudes between about 60 and 70 degrees north. All of these processes contribute to the Arctic carbon vault.

The pan-Arctic stockpiling of organic carbon has proceeded over geological timescales—some of this organic material is truly ancient and lies beyond the range of radiocarbon dating (~50,000 years). A good deal of the organic matter stored in Arctic permafrost comes from plant communities that flourished on the mammoth steppe tundra of the last ice age—an ecosystem for which there is no modern analogue. These plants extracted carbon from an ice age atmosphere. The deep permafrost is the best place on Earth for conserving plant and animal material. When the carcass of the Beresovka woolly mammoth was discovered in Siberia in 1901, beautifully preserved grasses and buttercups were lodged between its enormous teeth and its guts were stuffed with vegetation. That mammoth took its last meal over 40,000 years ago.

In 2012 scientists from the Russian Academy of Sciences successfully propagated fertile plants of *Silene stenophylla* from fruits that had been stored in the permafrost for over 30,000 years. This is the oldest plant material to be brought back to life. *Silene stenophylla* is a perennial herbaceous plant, a form of campion with white or light pink flowers that still grows on the eastern Siberian tundra. Ancient fruits were excavated from the fossil burrows of ground squirrels exposed in the steep banks of the Kolyma River. Some of the burrow chambers contained several hundred thousand fruits and seeds. All of this is very old carbon.

The permafrost carbon feedback

A cold Arctic with extensive permafrost is a very effective long-term carbon sink because the carbon is safely locked away as long as permafrost is maintained. However, soils come alive as the ground thaws; microbial communities flourish in the presence of liquid water, oxygen, and warmer temperatures, and the soft parts of long buried animals begin to rot. Under these conditions organic matter is broken down, and carbon is released to the atmosphere as carbon dioxide and methane (CH_4), another greenhouse gas.

While these processes have always taken place in the active layer during the Arctic summer, warmer summers are driving permafrost thaw to ever greater depths across vast regions. Permafrost is becoming fragmented and more sporadic. All of this is exposing increasing amounts of ancient organic matter to microbial attack. As Arctic ecosystems warm, some of the carbon losses will be offset by increased plant growth, peat formation, and the northward expansion of boreal forest. All of this will add carbon to the terrestrial pool. On balance, however, there is growing concern that the Arctic landscape has become a net supplier of greenhouse gases to the atmosphere. In other words, organic matter decomposition now exceeds accumulation. The release of CO_2 and CH_4 enhances warming, which leads to more extensive thawing of the permafrost and more outgassing. This is the permafrost–carbon-feedback loop. Carbon is also being lost to wildfires as dried out peatlands combust across the Arctic. Figure 29 shows multiple changes to the Arctic carbon budget under a warming climate.

Exploding ground

In 2014, a giant crater some 30 m in diameter and 70 m deep appeared suddenly on the Yamal Peninsula in north-west Siberia (Figure 30). Several others appeared that summer in the same area. These mysterious features had never been witnessed before. They have been attributed to the explosive release of methane gas that has accumulated at pressure in the sub-surface. This explanation is supported by the presence of ejected sediment found hundreds of metres from the craters. Russian scientists were alarmed to see features on this scale forming so rapidly. The Yamal Peninsula experienced unusually warm summers in 2012 and 2013, about 5°C above average. Air samples collected near the base of the crater showed very high concentrations of methane of about 9.6 per cent, when the atmosphere at ground level typically has a concentration of just 0.000179 per cent. These giant permafrost pockmarks eventually fill with water. Another crater appeared in late August during the very warm summer of 2020.

Thermokarst lakes

Thawing permafrost leads to the formation of lakes and ponds of various sizes. Thermokarst lakes are seasonally frozen bodies of water held in subsidence depressions created by the thawing of ground ice. The lake bottoms often sit on the permafrost surface. Satellite images show how these small lakes are densely packed

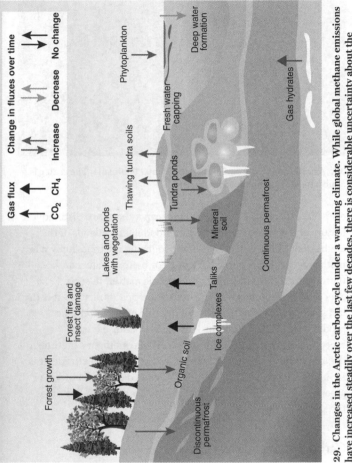

29. Changes in the Arctic carbon cycle under a warming climate. While global methane emissions have increased steadily over the last few decades, there is considerable uncertainty about the contribution from the Arctic.

129

30. A crater in the permafrost on the Yamal Peninsula. The crater is approximately 30 metres in diameter.

over vast expanses of the Arctic landscape; they are a distinctive feature of the great river deltas and floodplain wetlands of the High Arctic (Figure 31). Methane-producing microbes known as methanogens thrive in the oxygen-poor bottom waters of thermokarst lakes scavenging on organic matter. In such anaerobic environments, CH_4 production predominates—it is the last stage in the decomposition of organic matter. Methane production has attracted much attention because it is an exceptionally potent greenhouse gas. While methane's lifetime in the atmosphere is much shorter than that of carbon dioxide, it is far more efficient at trapping heat. Averaged over a century, the greenhouse contribution of CH_4 is more than twenty-five times the same amount of CO_2.

Carbon in Arctic rivers

During the summer thaw, soils can supply fine sediment and particulate organic carbon to Arctic rivers as hill slopes and river banks become unstable and the landscape is eroded by fluvial

31. Thermokarst lakes and braided channels on the Lena River delta. The image is approximately 180 km across.

processes. River flows also transport organic carbon in dissolved form. In light of concerns about permafrost degradation and changes to the Arctic carbon cycle, there is a need to better understand the riverine transport of both particulate and dissolved organic carbon. Arctic rivers transfer the solid and liquid products of thawing permafrost to the Arctic Ocean.

The sediments on the floor of the Arctic Ocean are rich in organic carbon. Because biological productivity in the Arctic Ocean is generally low—partly due to sea ice cover and limited light penetration—there is only a limited supply of *primary* marine organic carbon. The particulate organic matter we see in the sea floor sediments originates primarily from erosion of the surrounding landmasses and its supply is dominated by six giant Arctic river basins: the Mackenzie and Yukon in North America, and the Kolyma, Yenisey, Lena, and Ob in Siberia. These rivers carry very large loads of particulate organic carbon to the marine environment. At 1.8 million km², the Mackenzie River drains the largest catchment in North America after the Mississippi and carries the largest sediment load to the Arctic Ocean. Permafrost is present in various states over 80 per cent of its catchment area. Recent radiocarbon dating of the organic material transported by the Mackenzie yielded a mean radiocarbon age of 5,800 ± 800 years BP. This shows that old organic carbon is being exhumed from the Canadian permafrost and transported to the coastal zone. If this material is buried on large subsiding deltas, on the continental shelf, or in deeper basins, it can re-enter long-term storage. We need more information on the carbon budgets of these big rivers to better understand how they will respond to a warmer Arctic.

In comparison to most large rivers around the world, the great rivers of the Arctic contain the highest concentrations of dissolved organic carbon—another distinctive feature of the high northern latitudes. In the overall Arctic carbon budget, however, the loss of carbon from the terrestrial environment in this form is relatively small. Research in the Kolyma River basin of eastern Siberia—the world's largest catchment that is 100 per cent permafrost—shows that the transport of dissolved organic carbon is dominated by *modern* organic material that comes from strong leaching of the active layer and surface litter in early spring.

Methane beneath the sea

The East Siberian Arctic Shelf (ESAS) is the largest continental shelf in the world, covering an area of some 2 million km². It is roughly the size of Greenland and parts of the shelf extend offshore for more than a thousand kilometres. It is also the shallowest large continental shelf, with a mean water depth of just

50 m. During glacial stages of the Pleistocene, most of this shelf would have been well above sea level and part of the Arctic tundra landscape. This shelf is the reason why the remains of woolly mammoths and other megafauna have been found on Wrangel Island and the New Siberian Islands in the East Siberian Sea—it connected them to the mainland for most of the Pleistocene. When mammoths strode across this landscape it was part of the High Arctic permafrost zone. That permafrost still exists, but it now lies beneath the Arctic Ocean; its upper surface sits close to the modern seabed, a frozen relic of the Pleistocene ice age.

The frozen deposits of the East Siberian Arctic Shelf form the world's most extensive area of submarine permafrost. This hidden component of the cryosphere is currently attracting a great deal of scientific attention because it seals a vast reservoir of methane in the form of methane hydrate. Sometimes called methane ice, methane hydrate is a solid compound of methane gas and water that is flammable. The presence of Arctic Sea ice and the stability of the submarine permafrost have been critical factors in trapping the methane hydrate—like a cork in a champagne bottle. But sea ice retreat has allowed the shallow shelf waters to warm, and enhanced wave action encourages vertical mixing, so the seabed comes into contact with warmer surface waters. There is growing concern that sustained climate warming could free enough of this methane to rapidly accelerate global climate change via a positive feedback loop. It has been suggested that a rapid release of methane from ocean floor sediments triggered the Palaeocene–Eocene Thermal Maximum about 56 million years ago, when the Arctic region was much warmer than today, forested, and largely ice free.

Because the shelf waters are so shallow, methane bubbles can escape directly into the atmosphere, whereas in deeper waters most methane dissolves in the water column. Russian scientists have pioneered this research: a decade or so ago they reported the presence of anomalously high concentrations of dissolved methane in the bottom waters of the ESAS that pointed to *in situ* release from degrading submarine methane hydrate deposits. They have published images of methane plumes escaping from the seabed. Methane concentrations four times higher than normal have been observed in the atmosphere above the Eastern Siberian Sea, currently a global hotspot for methane venting. Methane emissions have been observed from other Arctic shelves too.

On a more optimistic note, a methane-consuming microbe has recently been discovered in Siberian upland soils. It has been found to extract methane from the atmosphere as a source of energy. Modelling studies by researchers at Purdue University suggest these methanotrophs may be offsetting some of the methane released from marine and lake environments.

Mammoths in the mud

The abundance of organic carbon in Arctic soils is not surprising when one considers the astonishing preservation of long-dead ice age animals recovered from the permafrost. Body parts and more or less complete carcasses of woolly mammoths have been discovered in the high latitudes for centuries. Important discoveries were made in the 19th century across the High Arctic by, for example, gold miners in the Yukon and reindeer herders in Yakutia. Natural processes of river bank collapse and mass movements on slopes have exhumed many animal carcasses but such discoveries are increasing in frequency as the permafrost thaws. The 21st century has also seen the advent of well-organized gangs of mammoth ivory hunters who use hydraulic mining techniques to prise tusks from the permafrost. Yakutia is a particular focus for this activity. A well-preserved set of woolly mammoth tusks can fetch thousands of dollars in China. The restrictions placed on the trade of elephant ivory have helped fuel the search for ancient ivory from long extinct mammoths. A by-product of this industry is a dramatic increase in the discovery of long-dead ice age animals in varying degrees of preservation (Figure 32).

32. The perfectly preserved head of an ice age wolf from the permafrost of Yakutia.

In summer 2018 Russian scientists discovered the head of an ice age wolf. This is a singular discovery given the remarkably well-preserved soft body parts including ears, tongue, and brain. All previous finds of this extinct species were of skulls and bones without soft tissues. Two cave lion cubs found in Yakutia in 2015 could be up to 50,000 years old—even their tiny whiskers were perfectly preserved in the frozen ground. Reindeer herders discovered the carcass of a cave bear on the New Siberian island of Bolshyoy Lyakhovsky in summer 2020. The cave bear became extinct about 20,000 years ago. DNA can be well preserved in the permafrost and these discoveries offer exciting prospects to study the evolutionary history of long extinct Pleistocene fauna.

Such remarkable discoveries only emphasize the worrying fact that the Arctic is warming and the landscape is changing. The permafrost zone has been Earth's largest terrestrial carbon sink for much of human history, but climate models predict severe and widespread degradation of the permafrost this century. The carbon trapped in Arctic permafrost is often called 'the sleeping giant of the global carbon cycle' because it is a massive pool of carbon that is not fully factored into climate change prediction models. The offshore methane hydrate reservoir is of particular concern. A deeper understanding of the impact of climate warming on the Arctic carbon cycle is of global importance.

Chapter 8
Arctic futures

The future of the Arctic and the wider world is being made in Svalbard. As we write, the Global Seed Vault had received over 1 million seed varieties from around the world. It was inaugurated in 2008 as an Arctic sanctuary to safeguard the genetic diversity of crops to help meet the UN goal of eliminating hunger by 2030. In May 2017, however, thawing permafrost and melting snow flooded the entrance tunnel, raising fears that even a concrete vault at 78° N, set deep in a mountainside less than 1,000 km from the pole, was not immune to the impacts of climate change. Seed deposits resumed in February 2020 after the Norwegian authorities spent around $10 million in emergency flood-proofing work and upgrades to the cooling system. The Norwegian Prime Minister Erna Solberg declared: 'It is the great insurance policy for food security' for the world's population.

The Svalbard archipelago is a hotspot of Arctic amplification. Very rapid warming has been keenly felt by this small community as it will challenge the ingenuity and resilience of all Arctic peoples. A warmer, wetter Arctic has led to new investment in housing and infrastructure in the Svalbard settlement of Longyearbyen. As the permafrost thaws, landslips, slumping, cratering, and ecological change follow and, more worryingly, ancient stores of carbon dioxide and methane are released into the atmosphere. As Svalbard's landscapes adjust to a new climate, the Norwegian authorities have authorized some radical changes to how they manage energy generation and mining on the archipelago. In January 2021, it was confirmed that the coal-fired heat and power plant in Longyearbyen will be shut down before 2025. This follows an earlier commitment to phase out the last Norwegian-owned working coal mine. In July 2020, when air temperatures exceeded 21°C, the coal mine was flooded by glacial meltwater.

While Norway is resigned to ending coal mining, Russia is showing no signs of making any changes. The Russian coal mine in the Svalbard settlement of Barentsburg is not due for closure. Norway suspects that Russia is in no hurry to end mining, however unprofitable and out of kilter it might be with the need to transition away from coal. If the mine were closed, the Russian settlement would be without an independent power source. Geopolitical considerations will continue to inform Russian decision-making, and even the Norwegian authorities will need to think carefully about the alternative energy options; either invest in local renewables or establish a sub-sea power cable between Svalbard and northern Norway. While the future is not coal, both Norway and Russia recognize that energy generation is integral to their continued occupation of Svalbard. The future of coal mining and energy generation in Svalbard is just one example of the dilemmas facing Arctic communities.

A greener Arctic goes hand in hand with greater fire risk. The summers of 2019 and 2020 saw wildfires rage across large swathes of the far north, especially in Russia, Canada, and Alaska. Strong winds deposited dark soot and charcoal on sea ice—increasing the likelihood of solar heat being absorbed rather than reflected. These feedbacks amplify environmental change. The fires—made worse in 2020 by record Arctic temperatures—belched huge amounts of carbon into the atmosphere. Some fires now burn through the winter as peat smoulders in the near-surface. When summer arrives, these zombie fires can reignite surface vegetation. The transformation of Arctic ecologies will have enduring consequences for all of us.

These environmental changes, however stark and widespread, will not dampen interest in economic development and strategic posturing. Arctic states and northern peoples are eager to improve social and economic conditions, but this can sit uneasily alongside demands to focus on environmental protection and climate change mitigation. Contradictions loom large in the Arctic. Svalbard coal currently generates electricity for the refrigeration units in the Global Seed Vault. Elsewhere, Arctic communities are under a range of pressures, and not all of them are rooted in climate change. Population decline is a long-term feature of the Russian Arctic, access to housing, food, and essential services is expensive and patchy in many parts of the Canadian north, poor health, suicide, and gender-based violence have been described as endemic in parts of Greenland. A more integrated and prosperous European Arctic is home to high-tech data centres, renewable energy projects, and more diverse social, economic, and cultural opportunities than typically found in parts of the Canadian and Russian north. Depending on where one looks in the Arctic, there is evidence of innovation and change but also social deprivation and continued reliance on mining and fossil fuel extraction. Many seasoned observers speak of multiple Arctics rather than one circumpolar region confronting common challenges, opportunities, and risks.

The Arctic continues to reveal how geopolitics and governance are never divorced from national economic development and strategic priorities. While indigenous peoples demand ever greater autonomy and self-determination across the north, Russian priorities will remain oil and gas extraction and strategic dominance in the waters north of the Federation. They are not alone. Norway continues to license hydrocarbon development in northern waters while pushing ahead with its plans to invest and develop the 'blue economy'. A more open Arctic, facilitated by sea ice shrinkage and global geopolitical interest, will place further pressures on Canada, Russia, and other Arctic states to invest in critical infrastructure along their northern flanks. Russia's Arctic strategies inevitably inform the perspectives of others. Norway worries that Svalbard might become a new geopolitical hotspot as it grapples with a post-coal future. The state of Alaska is wrestling with resource and environmental conflicts, with successive federal governments either imposing moratoriums on oil and gas development or seeking to license further development. US banks such as Morgan Stanley have pledged not to finance new oil and gas projects. And yet, Alaska is a resource-based economy, and the mining industry generated over $240 million to Alaskan native corporations in 2019 alone. Transitioning from the mining sector will not be straightforward and will be resisted by local settler and indigenous communities. Finland is contemplating reopening an iron ore mine in the north of the country as an employment generation strategy infused with appeals to mining heritage. Norway and Sweden have restarted iron ore mining in their northern territories, with similar employment-based justifications and supply-chain security concerns. While the Arctic has been a zone of peace for much for the last three decades, there is a great deal more discussion about possible military tension. Russia maintains its nuclear deterrent and substantial Northern Fleet on the Kola Peninsula, close to Norway, Finland, and Sweden. Reassuringly for Norway the Biden administration agreed to the dispatch of US B-1 bombers and military personnel to an air station in mid-Norway. The Arctic is hot property.

Compounding this sense of uncertainty is the fact that others such as the UK, China, and South Korea have been active in articulating their own wishes and interests in a global Arctic. This is not to claim that the Arctic was not globalized in the past. Resource extraction and pollution provide powerful examples of how the Arctic has been both harvested and dumped upon by others. We now worry about microplastics in sea ice rather than acid rain in Scandinavian lakes. Whatever the end-result, the Arctic remains vulnerable to far-travelled pollutants. Resource appraisals continue and workers from around the world construct and maintain essential Arctic infrastructure. Arctic states and third parties continue to assess the region's fish stocks, shipping prospects, and renewable energy potential. Governance structures such as the Arctic Council provide opportunities for both regional cooperation and global interaction. While the United States did warn in May 2019 that the Arctic was in danger of becoming caught up in global power struggles, it is important to bear in mind that any US administration is going to be eager to ensure that its own interests—resource, strategic, territorial—are protected.

The Arctic is still a haven of international peace and cooperation. The Arctic Council is often cited as a governance model that others could emulate. But the Arctic Council and other regional bodies are successful because the number of parties involved is small, the operating environment harsh, and the resource potential of the region remains concentrated in relatively few sectors and industries, where territorial ownership is not disputed. The greatest uncertainty comes from the impacts of climate change. Sea ice helps to keep the Arctic cold and plays a key role in moderating global climate, but Arctic amplification will continue to have ramifications for the rest of the world influencing extreme weather such as heatwaves, floods, and even heavy snowfall in the mid-latitudes. It carries significant costs and hazards for Arctic residents too. As resource pressures on land and in the ocean escalate with the impacts of global warming, such tensions come as no surprise.

Indigenous peoples are not going to simply watch and wait. They are mobilizing and will continue to insist upon their rights to make decisions about their futures. In November 2019, at the Arctic Leaders' Summit in Finland, delegates issued a declaration that 'nothing about us in the Arctic will happen without us as the indigenous peoples of the Arctic'. The Inuit Circumpolar Council pressurized the International Maritime Organization in London to introduce a ban on heavy fuels being used in Arctic shipping from 2024 onward, and even though Russia demanded an exemption until 2029, there will be a ban on such fuels in 2030. In coastal Alaska, native communities are mobilizing against the US federal government and petitioning the UN Rapporteurs to hear their case. The communities claim that their rights have been violated because insufficient attention was paid to the very real prospect of displacement due to rising sea levels. Indigenous activism is leading to new agreements over land and maritime resource planning in Canada and demands for 'climate justice'. Indigenous peoples had their land stolen by settler colonizers, their environments altered by activities such as forestry clearance and mining, and their cultural worlds turned upside down by dominant third parties. While it is now fashionable to speak of the disruptive consequences of contemporary climate change, the reality for indigenous peoples is that they have been living with external disruption for centuries.

Distinguishing between global geopolitical dynamics and local and regional realities in the Arctic is crucial. We need to be wary of assuming that the Arctic is a singular region. In many parts of the Arctic, there is an established pattern of collaboration and confidence-building in areas like the Barents Sea and Bering Strait. Even the remote Central Arctic Ocean will continue to have the benefit of a fisheries moratorium involving Arctic states and third parties such as China, South Korea, and the European Union. There are rules and frameworks such as the UN Law of the Sea Convention that help guide all parties, even if that same framework has encouraged Canada, Denmark/Greenland, and Russia to invest large sums of money mapping and surveying remote areas of the Arctic Ocean floor. No one thinks there are substantial subterranean resources to be harvested, but all three want to show their citizens and the wider world that their states extend furthest north. This tendency towards statecraft in the Arctic means that all parties will need to be watchful to ensure the region's reputation for peaceful coexistence remains intact.

References

In addition to the works listed in the Further Reading and online resources below, the following sources were particularly helpful as we researched this book. Many of the articles are open access. If access to any of the journal articles requires a subscription, email the corresponding author and ask for a PDF.

Chapter 1: The Arctic world

Evengård, B., et al. (eds) *The New Arctic*. Berlin: Springer (2016)

Hodgkins, R. Arctic breakdown: what climate change in the far north means for the rest of us. *The Conversation*, September (2019) <https://theconversation.com/arctic-breakdown-what-climate-change-in-the-far-north-means-for-the-rest-of-us-123309>

Holzworth, R. H., et al. Lightning in the Arctic. *Geophysical Research Letters* 48, e2020GL091366 (2021) <https://doi.org/10.1029/2020GL091366>

Lincoln, A., et al. (eds) *Arctic: Culture and Climate*. London: British Museum Press (2020)

Serreze, M. *Brave New Arctic: The Untold Story of the Melting North*. Princeton: Princeton University Press (2018)

Chapter 2: The physical environment

Batchelor, C. L., et al. The configuration of Northern Hemisphere ice sheets through the Quaternary. *Nature Communications* 10, 3713 (2019)

Ehlers, J. and Gibbard, P. L. The extent and chronology of Cenozoic Global Glaciation, *Quaternary International*, 164–5, 6–20 (2007) <https://doi.org/10.1016/j.quaint.2006.10.008>

Froese, D. G., et al. Ancient permafrost and a future, warmer Arctic. *Science* 321, 1648 (2006) <https://doi.org/10.1126/science.1157525>

Gardner, A., et al. Sharply increased mass loss from glaciers and ice caps in the Canadian Arctic Archipelago. *Nature* 473, 357–60 (2011) <https://doi.org/10.1038/nature10089>

The IMBIE Team, Shepherd, A., et al. Mass balance of the Greenland Ice Sheet from 1992 to 2018. *Nature* 579, 233–9 (2020). <https://doi.org/10.1038/s41586-019-1855-2>

Meredith, M., et al. Polar regions. In: *IPCC Special Report on the Ocean and Cryosphere in a Changing Climate*. Edited by Hans-Otto Pörtner et al. (2019) <https://www.ipcc.ch/srocc/>

NUNATARYUK Project (<https://nunataryuk.org/about>) *New map shows extent of permafrost in Northern Hemisphere* <https://news.grida.no/new-map-shows-extent-of-permafrost-in-northern-hemisphere>

Richter-Menge, J. and Druckenmiller, M. L. (eds) State of the climate in 2019: The Arctic. Special Online Supplement to the *Bulletin of the American Meteorological Society* 101 (2020) <https://doi.org/10.1175/BAMS-D-20-0086.1>

Spencer, A. M., et al. (eds) *Arctic Petroleum Geology*. London: Geological Society, Memoirs, 35 (2011)

Zhang, J., et al. Persistent shift of the Arctic polar vortex towards the Eurasian continent in recent decades. *Nature Climate Change* 6, 1094–9 (2016) <https://doi.org/10.1038/nclimate3136>

Zimov, S. A., et al. Permafrost and the global carbon budget. *Science* 312, 1612–13 (2006) <https://doi.org/10.1126/science.1128908>

Chapter 3: Arctic ecosystems

Anderson, R. Welcome to Pleistocene Park. *The Atlantic*, April 2017 <https://www.theatlantic.com/magazine/archive/2017/04/pleistocene-park/517779/>

Arctic Biodiversity Assessment Report (2013) CAFF (Conservation of Arctic Flora and Fauna) <https://www.arcticbiodiversity.is/index.php/the-report>

Berner, L. T., et al. Tundra plant above-ground biomass and shrub dominance mapped across the North Slope of Alaska.

Environmental Research Letters 13, 035002 (2018) <https://doi.org/10.1088/1748-9326/aaaa9a>

Billings, W. D. Constraints to plant growth, reproduction, and establishment in Arctic environments. *Arctic and Alpine Research* 19, 357–65 (1987) <https://doi.org/10.1080/00040851.1987.12002616>

Davidson, S. C., et al. (2020) Ecological insights from three decades of animal movement tracking across a changing Arctic. *Science*, 370 (6517), 712–15, DOI: 10.1126/science.abb7080

Frost, G. V., et al. Tundra Greenness. *Arctic Report Card: Update for 2019* <https://arctic.noaa.gov/Report-Card/Report-Card-2019/ArtMID/7916/ArticleID/838/Tundra-Greenness>

Fuglei, E. and Tarroux, A. Arctic fox dispersal from Svalbard to Canada: one female's long run across sea ice. *Polar Research* 38 (2019) <https://doi.org/10.33265/polar.v38.3512>

Gill, J. L., et al. Pleistocene megafaunal collapse, novel plant communities, and enhanced fire regimes in North America. *Science* 326, 1100–3 (2009) <https://doi.org/10.1126/science.1179504>

Kovacs, K. M., et al. The endangered Spitsbergen bowhead whales' secrets revealed after hundreds of years in hiding. *Biology Letters* 16, 20200148 (2020) <https://doi.org/10.1098/rsbl.2020.0148>

Meltofte, H. (ed.) *Arctic Biodiversity Assessment 2013: Status and Trends in Arctic Biodiversity.* Conservation of Arctic Flora and Fauna (2013) <https://www.caff.is/assessment-series/arctic-biodiversity-assessment/233-arctic-biodiversity-assessment-2013>

Selås, V. Timing of population peaks of Norway lemming in relation to atmospheric pressure: A hypothesis to explain the spatial synchrony. *Nature Scientific Reports* 6, 27225 (2016) <https://doi.org/10.1038/srep27225>

Walker, D. A., et al. The Circumpolar Arctic vegetation map. *Journal of Vegetation Science* 16, 267–82 (2005) <https://doi.org/10.1111/j.1654-1103.2005.tb02365.x>

Yong, E. Narlugas are real. *The Atlantic*, 20 June (2019) <https://www.theatlantic.com/science/archive/2019/06/narluga-very-strange-hybrid-whale/592057/>

Yu, Q., et al. Circumpolar arctic tundra biomass and productivity dynamics in response to projected climate change and herbivory. *Global Change Biology* 23, 3895–907 (2017) <https://doi.org/10.1111/gcb.13632>

Chapter 4: Peoples of the Arctic

Arctic Indigenous Peoples: <https://www.arcticcentre.org/EN/arcticregion/Arctic-Indigenous-Peoples>

The Canadian Encyclopaedia of Indigenous Peoples: <https://www.thecanadianencyclopedia.ca/en/timeline/first-nations>

Heleniak, T. Migration in the Arctic. *Arctic Yearbook 2014* <https://arcticyearbook.com/images/yearbook/2014/Scholarly_Papers/4.Heleniak.pdf>

Pitulko, V., et al. Early human presence in the Arctic: Evidence from 45,000-year-old mammoth remains. *Science* 351, 260–3 (2016) <https://doi.org/10.1126/science.aad0554>

Vogel, B. and Bullock, R. Institutions, indigenous peoples, and climate change adaptation in the Canadian Arctic. *GeoJournal* (2020) <https://doi.org/10.1007/s10708-020-10212-5>

Chapter 5: Exploration and exploitation

Arctic Deeply *Economic Development and the Future of Mining in the Canadian North* (2016) <https://deeply.thenewhumanitarian.org/arctic/community/2016/08/16/economic-development-and-the-future-of-mining-in-the-canadian-north>

Cone, M. *Silent Snow: The Slow Poisoning of the Arctic*. New York: Grove Press (2006)

Morgunova, M. Why is exploitation of Arctic offshore oil and natural gas resources ongoing? A multi-level perspective on the cases of Norway and Russia. *Polar Journal* 10, 64–81 (2020) <https://doi.org/10.1080/2154896X.2020.1757823>

Robinson, M. *The Coldest Crucible*. Chicago: University of Chicago Press (2006)

Royal Museums Greenwich: *Hudson Bay Company trading token— one Made Beaver:* <https://collections.rmg.co.uk/collections/objects/40198.html>

Potter, R. *Visions of the North* <https://visionsnorth.blogspot.com/p/arctic-exploration-brief-history-of.html>

Sámi Arctic Strategy (2019) <https://www.saamicouncil.net/documentarchive/the-smi-arctic-strategy-samisk-strategi-for-arktiske-saker-smi-rktala-igumuat>

Wright, J. Open polar sea. *Geographical Review* 43, 338–65 (1953) <https://doi.org/10.2307/211752>

Chapter 6: Arctic governance

Arctic Council <https://arctic-council.org/en/>

Koivurova, T. Is this the end of the Arctic Council and Arctic governance as we know it? *High North News* 12 December (2019) <https://www.highnorthnews.com/en/end-arctic-council-and-arctic-governance-we-know-it>

Østhagen, A. *The Good, the Bad and the Ugly: Three Levels of Arctic Geopolitics.* University of Waterloo Balsillie Papers 14 December (2020) <https://balsilliepapers.ca/bsia-paper/the-good-the-bad-and-the-ugly-three-levels-of-arctic-geopolitics/>

Steinveg, B. *Governance by Conference? Actors and Agendas in Arctic Politics.* PhD thesis, The Arctic University of Norway (2021)

Vincent, J. Mark Zuckerberg shares pictures from Facebook's cold, cold data center. *The Verge,* 29 September (2016) <https://www.theverge.com/2016/9/29/13103982/facebook-arctic-data-center-sweden-photos>

Young, O. R. Is it time for a reset in Arctic governance? *Sustainability* 11, 4497 (2019) <https://doi.org/10.3390/su11164497>

Chapter 7: The Arctic carbon vault

Arctic Climate Impact Assessment (ACIA) *Impacts of a Warming Arctic.* Cambridge: Cambridge University Press (2004) <https://www.amap.no/documents/doc/impacts-of-a-warming-arctic-2004/786>

Brouillette, M. How microbes in permafrost could trigger a massive carbon bomb. *Nature* 591, 360–2 (2021) <https://doi.org/10.1038/d41586-021-00659-y>

Chuvilin, E., et al. New catastrophic gas blowout and giant crater on the Yamal Peninsula in 2020: Results of the expedition and data processing. *Geosciences* 11 (2021) <https://doi.org/10.3390/geosciences11020071>

Hilton, R., et al. Erosion of organic carbon in the Arctic as a geological carbon dioxide sink. *Nature* 524, 84–7 (2015) <https://doi.org/10.1038/nature14653>

Shakhova, N., et al. Extensive methane venting to the atmosphere from sediments of the East Siberian Arctic Shelf. *Science* 327, 1246–50 (2010) <https://doi.org/10.1126/science.1182221>

Yashina, S., et al. Regeneration of whole fertile plants from 30,000-y-old fruit tissue buried in Siberian permafrost. *PNAS* 109 (10) 4008–13 (2012) <https://doi.org/10.1073/pnas.1118386109>

Chapter 8: Arctic futures

Gertner, J. *The Ice at the End of the World: An Epic Journey into Greenland's Buried Past and Our Perilous Future.* New York: Penguin (2019)

Struzik, E. *Future Arctic.* Washington DC: Island Press (2018)

The Svalbard Global Seed Vault <https://www.croptrust.org/our-work/svalbard-global-seed-vault/>

Wormbs, N. (ed.) *Competing Arctic Futures: Historical and Contemporary Perspectives.* London: Palgrave (2018)

The Arctic on the world wide web

The following selection is intended to give a sense of the richness and diversity of online learning resources addressing the Arctic.

The Arctic environment

For a UK-sponsored website 'Discovering the Arctic' with support from the Royal Geographical Society, NERC Arctic Office and the British Antarctic Survey amongst others: <https://discoveringth-earctic.org.uk/>

The British Museum's Arctic Culture and Climate exhibition (2020–1) yielded an informative and richly illustrated web-based resource: <https://www.britishmuseum.org/exhibitions/arctic-culture-and-climate>

The National Snow and Ice Data Center is widely regarded as one of the most authoritative sources for information about the Arctic cryosphere: <https://nsidc.org/>

'Chasing Ice' is the story of environmental photographer James Balog's mission to capture a multi-year record of the Arctic's changing glaciers <https://chasingice.com/>

The Arctic Great Rivers Observatory stores important data on the fluvial systems draining permafrost landscapes of Siberia and North America <https://arcticgreatrivers.org/>

On Russia's approach to the Arctic with official support from the Russian Geographical Society: <https://arctic.ru/>

University of Colorado Boulder has an excellent educational resource on Arctic and polar learning resources: <https://mosaic.colorado.edu/remote-online-and-home-polar-learning-resources>

Universities in the north have been proactive in archiving and curating cultural, historical, and economic resources of the circumpolar Arctic. A good example is Athabasca University: <https://www.athabascau.ca/indigenous/resources/research-library/inuit-northern-people.html>

Michael Robinson's podcast series on Arctic science, history, and exploration is highly recommended: <https://timetoeatthe-dogs.com/>

Indigenous peoples

The Arctic Council's Indigenous Peoples Secretariat: <https://arctic-council.org/en/about/indigenous-peoples-secretariat/>

The Coat of Arms of Nunavut explained https://assembly.nu.ca/about-legislative-assembly/coat-arms-nunavut

For a story-map approach to the 'Indigenous peoples of the Arctic' produced by the GRID-Arendal non-profit environment communication centre: <https://www.arcgis.com/apps/Cascade/index.html?appid=2228ac6bf45a4cebafc1c3002ffef0c4>

A good example of indigenous advocacy and activism is the Bering Sea Elders Group in Alaska with plenty of examples of how social media and video is being used to highlight the human costs of climate change in the Arctic: <http://www.beringseaelders.org/>

Politics, resources, and governance

The Arctic Council, the main intergovernmental forum for Arctic governance, maintains a wealth of resources and documentary links: <https://arctic-council.org/en/> and there are links to the six main permanent participants such as the Inuit Circumpolar Council, Russian Association of Indigenous Peoples of the North, and Sámi Council.

For details on Arctic resources see this weblink maintained by the European Environment Agency: <https://www.eea.europa.eu/data-and-maps/figures/arctic-resources>

Nordregio is a Nordic research centre for regional development and planning, established by the Nordic Council of Ministers, which maintains a useful inventory of Nordic Arctic development: <https://nordregio.org/research-topics/arctic-issues/#>

Official website of the government of Greenland: <https://naalakkersuisut.gl/en/Naalakkersuisut>

Eric Paglia's Polar Geopolitics podcast is also a highly recommended source of insights into Arctic geopolitics and governance: <http://www.polargeopolitics.com/>

Further reading

Buchanan, E. *Red Arctic? Russian Strategy under Putin*. New York: Brookings Institute Press (2021)

Cameron, E. *Far Off Metal River: Inuit Lands, Settler Stories, and the Making of the Contemporary Arctic*. Vancouver: UBC Press (2016)

Cruikshank, J. *Life Lived Like a Story: Life Stories of Three Yukon Native Elders*. Vancouver: UBC Press (1990)

Depledge, D. *Britain and the Arctic*. London: Palgrave (2018)

Durfee, M. and Johnstone, R. L. *Arctic Governance in a Changing World*. Lanham, Md: Rowman and Littlefield (2020)

Friesen, T. M. and Mason, O. K. (eds) *The Oxford Handbook of the Prehistoric Arctic*. Oxford: Oxford University Press (2016)

Gjørv, G. H., Lanteigne, M., and Sam-Aggrey, H. (eds) *Routledge Handbook of Arctic Security*. London: Routledge (2020)

Hønneland, G. *Russia and the Arctic: Environment, Identity and Foreign Policy*. London: I B Tauris (2019)

IPCC Sixth Assessment Report 2021. https://www.ipcc.ch/report/ar6/wg1/

Jakobsen, K. and Frank, S. *Arctic Archives—Ice, Memory, and Entropy*. Berlin: Verlag (2021)

McCannon, J. *Red Arctic: Polar Exploration and the Myth of the North in the Soviet Union, 1932–1939*. Oxford: Oxford University Press (1997)

McCannon, J. *A History of the Arctic*. London: Reaktion (2012)

McCorristine, S. *The Spectral Arctic: A History of Dreams and Ghosts in Polar Exploration*. London: UCL Press (2018)

Menezes, D. W. and Nicol, H. N. *The North American Arctic: Themes in Regional Security*. London: UCL Press (2019)

Nuttall, M. *Climate, Society and Subsurface Politics in Greenland: Under the Great Ice.* London: Routledge (2017)

Routledge, K. *Do You See Ice? Inuit and Americans at Home and Away.* Chicago: University of Chicago Press (2018)

Sakakibara, C. *Whale Snow: Iñupiat, Climate Change, and Multispecies Resilience in Arctic Alaska.* Tempe, Ariz.: University of Arizona Press (2020)

Thomas, D. N., et al. *The Biology of Polar Regions* (2nd edn) Oxford: Oxford University Press (2008)

Vassnes, B. *Kingdom of Frost: How the Cryosphere Shapes Life on Earth.* Vancouver: Greystone Books (2020)

Vincent, W. F. and Laybourn-Parry, J. (eds) *Polar Lakes and Rivers.* Oxford: Oxford University Press (2008)

Wadhams, P. *A Farewell to Ice.* London: Allen Lane (2016)

Wilson Rowe, E. *Arctic Governance: Power in Cross-border Cooperation.* Cambridge: Cambridge University Press (2018)

Woodward, J. C. *The Ice Age: A Very Short Introduction.* Oxford: Oxford University Press (2014)

Woon, C. Y. and Dodds, K. (eds) *Observing the Arctic: Asia in the Arctic Council and Beyond.* Cheltenham: Edward Elgar (2020)

Index

Arctic haze 112
Arctic homelands 75
Arctic Indigenous peoples 63, 72–78, 120
Arctic Inspiration Prize 67
Arctic Leaders' Summit 2019 141
Arctic migrations 68–72
Arctic Marine Shipping Assessment (2009) 123
Arctic Monitoring and Assessment Programme (AMAP) 112
Arctic National Wildlife Refuge (ANWR) 76
Arctic Ocean (*see also* Central Arctic Ocean) 110, 142
  bathymetry 18–21
  currents 17, 36–38
  dimensions 18
  gateways 21
  salinity 20, 37
  sea ice (*see* sea ice)
  stratification 20, 24, 37
  water masses 20, 37, 38
Arctic paradox 95
Arctic poppy (*Papaver radicatum*) 41
Arctic Report Card 54, 75, 117
Arctic states 63–65, 78–79, 84, 139
Arctic tern (*Sterna paradisaea*) 46
Arkhangelsk 103
arktos 3
arsenic 101
Athabascan 9, 63, 76
Atlantic bluefin tuna (*Thunnus thynnus*) 58
Atlantic Meridional Overturning Circulation (AMOC) 37
Atlantic Ocean 20–21, 37–38, 58, 85, 86, 117, 119
Atlanticification 38
auks 43
aurora borealis 11
Australia 100, 120
Axel Heiberg Island 19, 33

B

Baffin Island 6, 33, 85
Barentsburg 94, 138
Barents Kara ice sheet 35
Barents Sea 33, 35, 37, 38, 58, 59, 81, 105, 142
Barents shelf 18, 21
Barnes Ice Cap 33
beavers (*see* North American beaver)
beaver exploitation 89–92
beaver hats 91
beluga whale (*Delphinapterus leucas*) 52–53
Bennett, James Gordon 88
Bering Land Bridge 19, 39, 68–70
Bering Sea 21, 60, 75
Bering Sea Elders Group 75
Bering Strait 21, 69, 75, 88, 142
Beringia 42, 69
biodiversity 43, 52
biomass 47, 53–55, 58, 107, 125
biome 39, 41, 42, 46, 47, 50, 53, 54, 57, 58
birds 10, 43, 44 (*also listed by species*)
bison 47, 57, 68
black carbon 76
blizzards 18, 62
blubber (*see also* whaling) 95
borders 70, 76, 116
boreal forest (*see also* forests and forestry, taiga) 31, 41, 127
boreal winter (*see also* winter) 16, 22
bowhead whale (*Balaena mysticetus*) 61
braided river (*see also* rivers) 33, 131
Bristol 86
British-Irish ice sheet 35
brown lemming (*Lemmus trimucronatus*) 50
Bylot Island, Nunavut Territory 51

Index

LANDSCAPES AND GEOMORPHOLOGY
A Very Short Introduction
Andrew Goudie & Heather Viles

Landscapes are all around us, but most of us know very little about how they have developed, what goes on in them, and how they react to changing climates, tectonics and human activities. Examining what landscape is, and how we use a range of ideas and techniques to study it, Andrew Goudie and Heather Viles demonstrate how geomorphologists have built on classic methods pioneered by some great 19th century scientists to examine our Earth. Using examples from around the world, including New Zealand, the Tibetan Plateau, and the deserts of the Middle East, they examine some of the key controls on landscape today such as tectonics and climate, as well as humans and the living world.

www.oup.com/vsi

GLOBALIZATION
A Very Short Introduction
Manfred Steger

'Globalization' has become one of the defining buzzwords of our time - a term that describes a variety of accelerating economic, political, cultural, ideological, and environmental processes that are rapidly altering our experience of the world. It is by its nature a dynamic topic - and this *Very Short Introduction* has been fully updated for 2009, to include developments in global politics, the impact of terrorism, and environmental issues. Presenting globalization in accessible language as a multifaceted process encompassing global, regional, and local aspects of social life, Manfred B. Steger looks at its causes and effects, examines whether it is a new phenomenon, and explores the question of whether, ultimately, globalization is a good or a bad thing.

GEOPOLITICS
A Very Short Introduction
Klaus Dodds

In certain places such as Iraq or Lebanon, moving a few feet either side of a territorial boundary can be a matter of life or death, dramatically highlighting the connections between place and politics. For a country's location and size as well as its sovereignty and resources all affect how the people that live there understand and interact with the wider world. Using wide-ranging examples, from historical maps to James Bond films and the rhetoric of political leaders like Churchill and George W. Bush, this Very Short Introduction shows why, for a full understanding of contemporary global politics, it is not just smart - it is essential - to be geopolitical.

'Engrossing study of a complex topic.'

Mick Herron, Geographical.

GEOGRAPHY
A Very Short Introduction
John A. Matthews & David T. Herbert

Modern Geography has come a long way from its historical roots in exploring foreign lands, and simply mapping and naming the regions of the world. Spanning both physical and human Geography, the discipline today is unique as a subject which can bridge the divide between the sciences and the humanities, and between the environment and our society. Using wide-ranging examples from global warming and oil, to urbanization and ethnicity, this *Very Short Introduction* paints a broad picture of the current state of Geography, its subject matter, concepts and methods, and its strengths and controversies. The book's conclusion is no less than a manifesto for Geography' future.

> 'Matthews and Herbert's book is written- as befits the VSI series- in an accessible prose style and is peppered with attractive and understandable images, graphs and tables.'
>
> Geographical.

www.oup.com/vsi